Integrating Information Technology into Education

Integrating Information Technology into Education

Edited by
Deryn Watson
and
David Tinsley

Published by Chapman & Hall on behalf of the
International Federation for Information Processing (IFIP)

CHAPMAN & HALL

London · Glasgow · Weinheim · New York · Tokyo · Melbourne · Madras

Published by Chapman & Hall, 2–6 Boundary Row, London SE1 8HN, UK

Chapman & Hall, 2–6 Boundary Row, London SE1 8HN, UK

Blackie Academic & Professional, Wester Cleddens Road, Bishopbriggs, Glasgow G64 2NZ, UK

Chapman & Hall GmbH, Pappelallee 3, 69469 Weinheim, Germany

Chapman & Hall USA, One Penn Plaza, 41st Floor, New York NY 10119, USA

Chapman & Hall Japan, ITP-Japan, Kyowa Building, 3F, 2-2-1 Hirakawacho, Chiyoda-ku, Tokyo 102, Japan

Chapman & Hall Australia, Thomas Nelson Australia, 102 Dodds Street, South Melbourne, Victoria 3205, Australia

Chapman & Hall India, R. Seshadri, 32 Second Main Road, CIT East, Madras 600 035, India

First edition 1995

© 1995 IFIP

Printed in Great Britain by TJ Press Ltd, Padstow, Cornwall

ISBN 0 412 62250 5

A catalogue record for this book is available from the British Library

∞ Printed on permanent acid-free text paper, manufactured in accordance with ANSI/NISO Z39.48-1992 and ANSI/NISO Z39.48-1984 (Permanence of Paper).

CONTENTS

Preface

Deryn Watson and David Tinsley

The topic of the conference, integrating information technology into education, is both broad and multi-facetted. In order to help focus the papers and discussion we identified 7 themes:

- Current developments in society and education influencing integration;
- Teachers, their roles and concerns;
- Learners, their expectations of and behaviour in an integrated environment;
- Developments and concerns in the curriculum;
- Successes and failures in existing practice;
- Organisation and management of integrated environments;
- Identification of social and political influences.

Each author was invited to focus on one theme, and these remained strands throughout as can be seen from the short papers and focus group reports.

The first and most significant concern therefore was to be clear about our notions of integration; what do we mean and how is this relevant? Our keynote paper from Cornu clearly marked out this debate by examining the notion of integration and alerting us to the fact that as long as the use of IT is still added to the curriculum, then integration has not yet begun. Full integration, that is the evolution of new concepts and change in the curriculum to reflect a full incorporation of new technologies, is still far from being perceived and understood, let alone achieved. For there to be progress there will need to be change at all levels, and further reflection about the nature of the education system of tomorrow. The current situation of juxtaposition has to shift towards integration before we can hope to achieve the level of "invisibility", because IT is no longer special, such as that achieved by the telephone and television.

The interpretation of this aim, and potential methods for achieving it, was one of the threads that ran through the papers and at the conference.

Duchateau bemoans the fact that the computer is often being tacked onto an educational system that in itself is inappropriately designed for current society; a Socratic model with the focus on the teacher inhibits a clear perspective of the role of IT in changing the experience and indeed notion of the learner. The computer may however be seen as a mirror to help teachers reflect and revise their notions of their role.

The next four papers each address the roles and needs of teachers. Hodgson focussed on the key aspect of the teachers pedagogical agenda, and is concerned at the new range of competencies that teachers need to embrace. These cover both their subject specialism, their understanding of informatics, and their overall notions of being a teacher. This makes teacher education a key issue. Nicassio considers that an important aspect has to be the concerns teachers already express which are inhibiting integration, which include barriers to full access, a lack of personal preparedness and the problem of unanticipated negative consequences of computer use. He proposes that an action research role for teacher has potential for removing these barriers.

Olson has a similar theme in that he identifies a range of teachers' concerns on the difficulty of achieving interaction with existing practice, and raises the problem that teachers are not provided with enough time to reflect on practice and change. Indeed teachers appear to be expected to take a substantial risk in accommodating IT into their teaching that threatens their classroom ethos. Like Duchateau, he notes the value of computers in that they may help teacher to confront their experience; change is not just adoption and integration of the technology, but involves personal reflection and adaptation as well. Ridgeway and Passey provide further perspective on this theme by also focussing on the approaches to change. The advocates of computers, through over ambitious claims and an underestimation of practical constraints, seed failure; they challenge teachers' fundamental values and practices. They propose a means of conceptualising the change by identifying stages in the process that addresses the needs as perceived by the teachers.

The next two papers take us away from the role of the teacher and remind us of learners and their experiences. Johnson states that despite the rhetoric in some publications, most learners do not receive a consistent experience of using IT in their subject based curriculum. Indeed the amount of access to IT they have is patchy and below a threshold level that appears to make any impact possible. Although he accepts that the role of the teacher is of importance, the effect of the teacher is limited while access and curriculum incorporation is so weak. Nicholson proposes that rather than use a technocentric model of the curriculum for integration, a more humanistic model should prevail with the focus on the student as the constructor of knowledge, charting a variety of independent and flexible routes through traditional curriculum areas through the facilitation of computational media.

Two papers follow which are fulfil the demands of Cornu in that they discuss the role of software in changing the conceptual basis of a subject, in both cases mathematics. The Labordes argue the case for Cabri-géomètre which provides an environment for the exploration of new problems, and in particular in relation to way mathematical objects may now be perceived by learners through novel manipulation. Kobayashi describes the way students, through moving schema in real time, may make comparisons between an object before and after transformation, and then focus on invariants. The reality of the achievements presented by both these papers provides a heartening window into the possibilities of change.

The debate in the next three papers moves on to an issue that has often exercised IFIP meetings, that is the role of informatics itself as a subject of study. Van Weert argues for a new informatics literacy, based on discipline integrated informatics; he identifies modelling and team work as key elements of informatics itself which has a wider role in all disciplines. He proposes that this aspect of integrated informatics in disciplines will be separate from the application orientated informatics that will be an elective and specialist form of study. Obershelp in contrast argues for obligatory informatics courses, that focus on the development of algorithmic and analytical competency for the learner. He states that in no other field can fundamental protocols such as parallelisms be handled. Graf opens a new dimension by arguing for a study if the history of informatics as a means of developing intercultural studies. His examples of hexagrams from the I Ching and Chinese calculating machines from the seventeenth century provide a delightful alternative perspective and role for informatics in education.

The next five papers return us to the reality of integration by their reports from widely different environments of the situation as it is today; all use classroom based studies as a basis for their comments. Abas reminds us that many countries need to focus on computer literacy projects, and that schools often face, particularly in rural areas, problems of scarcity of resources as basic as a reliable electricity supply. Despite such problems, she identifies the role of teacher education and preparation as key factors in achieving success. Ruiz and Quintana identify another set of factors by reporting on the role of the internal factors to a school which impinge on the quality of integration. These appear to include the schools' internal decision making processes, the pedagogical perceptions of access to the resource by the students and the participation levels of the teachers. This paper echoes aspects of Johnson's, but they focus clearly on the partnership between the individuals and the institution that is required.

Ween identifies exactly the same two major aspects, school and individual teacher level, but concludes that the teacher is the most dominant. His statements about the importance of teachers beliefs with respect to their subject and pedagogy remind us of the earlier papers by Hodgson and Olson. But he also stresses the problems they face if they are working in a non-collaborative school culture. The focus of Hurst's paper is the range of barriers, or windmills to be fought, if a subject specialist teacher is to achieve appropriate use of IT. From the headmaster to the computer specialist and subject colleagues, the battles that had to be fought clearly echoed problems of an isolated innovator. But she proposes that as colleagues turn into allies, the culture of cooperation can result in change. Chen focusses not just on the role of the teacher, and particularly the style of instructional guidance, but reminds us of the key perception of the learner. The learning characteristics of the students were explored in relation to preferred styles in association with tutorial help, and she suggests that the interplay between these two variables would appear to be important.

In the next two papers, these studies based on individual classrooms are placed in the broader context of national policies. From the experience of national projects in Spain, San Jose identifies elements that have not appeared before, in particular the role of IT in special needs and also the role of IT in fostering new cross-diciplinary perspectives. He reminds us that costs are still a major inhibiting factor and as a result places great emphasis on the use of general tools and multimedia as trends likely to have most impact on teaching and learning in the near future. Sakamoto and Miyashita identify the importance of government initiatives in spearheading and maintaining a momentum of implementation. Never-the-less they recognise the problems faced at the school level; they remind us that some teachers gain a sense of achievement in using IT that may be both trivial and inappropriate.

Teague and Clarke remind us that informatics still appeals more to males than females; boys consistently use computers more than girls in the home, school and community. They point to the importance of school as a location for providing a range of experiences to all pupils that may encourage a change of perception; in particular they identify the opportunity for informal use as a key factor for girls. Griffith, Murphy and Diaz however point to a further disturbing issue. Their study clearly showed that students' expectations of an integrated informatics society was formed principally by futurist portrayals in television and film. Their understandings were mediated substantially by social class. Working class and minority students are more likely to see the social contradictions of information technology. But it was Griffith who provided us with a timely reminder of the main issue of the conference. When presenting her paper on the last day, she asked how we could hope to have such lofty ideals as those espoused by Cornu when we are living in a world that has severe problems with integration within society.

Collis finally reminds us of the rapid changes in society and its perceptions of the role of information technologies; the current vision relates to a new world of an "information superhighway". The relationship between these political and economic forces, and the professional education community, may be the area in which integration is most likely to be achievable. Networking taps human needs to communicate and she argues for the adoption of a perception of community networking of which education is a part.

With this last paper we therefore see a direct strand back to the first. The role of integration, according to Cornu, is but a step in the path towards the generalisation of IT into everyday curriculum and teaching concepts. The subsequent papers reflect the tensions and issues that such a goal generates. The aim of the whole collection, including the short papers and focus group reports, is to provide a focus for further debate and forward action on integrating information technology into education.

Deryn Watson is a Senior Lecturer in Education at the Centre for Educational Studies, King's College London. Since the mid 1970's she has been actively involved in the research and development of computer assisted learning materials in the Humanities and Languages, with concerns for both models of software development and the potential for interactive learning to stimulate the exercise and development of process skills amongst learners in the classroom. She lectures on the role of IT in schools and has recently completed research into the impact of IT on children's achievements.

David Tinsley retired from his post as Chief Inspector of Training in 1992 and has recently joined the Red Cross as a manager. His early career was in teaching, during which time he helped IBM to develop the first computer for schools. After working for the National Computing Centre he was a General Inspector of Schools in Birmingham and then Education Officer for Further and Higher Education. He joined the Government Service in 1984 as Director of the Open Tech Programme and then pioneered a national inspection service for the Youth Training Scheme. David has been a member of IFIP WG 3.1 for over 25 years and has edited or contributed to a number of international guidelines during that time.

The IFIP working conference in Barcelona

This book derives from the IFIP Working Group 3.1 working conference "Integrating Information Technology into Education" in Barcelona, October 1994. The conference took a forward look at the issue of integration of information technology, and related problems of quality, at the secondary education level (students aged 11 - 18 years). It was the result of a partnership between the International Federation of Information Processing and the Department of Education of the Generalitat de Catalunya, Spain, based on shared interests in the development and integration of information technology into education. The programme and organisation of the working conference fell under the responsibility of the Programme committee, the Organising Committee and the Conference Secretariat.

Programme Committee

Bernard Cornu (chair), IUFM Grenoble, France
Klaus-D Graf (short papers), Free University Berlin, Germany
Bernard Hodgson (focus groups), Laval University, Quebec, Canada
Raymond Morel (focus groups), CIP Geneva, Switzerland
Ferran Ruiz Tarrago (organizing committee chair),
      Generalitat de Catalunya, Spain
Harriet Taylor (short papers), Louisiana State University, USA
David Tinsley (managing editor), United Kingdom
Deryn Watson (technical editor), King's College London, United Kingdom
Tom van Weert (vice-chair), University of Nijmegen, The Netherlands

Organising Committee
from the Generalitat de Catalunya, Departament d'Ensenyament

Francesc Busquets	Xon Camacho	Jordi Castells
Celia Fransitorra	Marian Minguillon	Ferran Ruiz Tarrago (chair)
Jordi Vivancos (vice -chair)		

Conference Secretariat
from OTAC, Barcelona

Josep Lluis Laborda	Elvira Martinez	Jesus Monforte

IFIP and Working Group 3.1

The International Federation for Information Processing (IFIP) is a multinational federation of professional and technical organisations concerned with information processing, founded in 1960 under the auspices of UNESCO. IFIP is dedicated to improve worldwide understanding about the role information processing can play in all walks of life, and to increase communication among practitioners of all nations. Members of IFIP are national organisations in the field of information processing.

Technical work, which is at the heart of IFIP's activities, is managed by a series of Technical Committees (TC). Each Technical Committee is composed of representatives of IFIP Member organisations. Technical Committee 3 is on Education. Under each technical committee there operate working groups which consist of specialists who are individually appointed by their peers independent of nationality. Working Group 3.1 (WG 3.1) is one such group which has its focus on secondary education.

IFIP WG3.1 has a tradition of organising working conferences every two years or so. The more recent ones were: Informatics and the teaching of mathematics, Sofia, Bulgaria (1987); Educational software at the secondary level, Reykjavik, Iceland (1989); and The impact of informatics on the organisation of education, Santa Barbara California, USA, (1991). These conferences have special organisational features: they are small with invited attendees; a series of invited papers are presented to stimulate discussion; other short papers are presented; and focus groups are formed from all participants which generate an exchange of ideas on particular topics and present at the end of the conference position papers reflecting their debate. These special features allow in-depth treatment of the conference topic, informal exchange of ideas and opinions, and encourage the establishment of new professional relationships.

The aim of this book is to publish the invited papers and to reflect the debate that occured during the conference through the short papers and focus groups.

The Generalitat de Catalunya

Catalonia is in the north-east of the Iberian Peninsula. Its language and socio-cultural, historical and political traditions make it quite different from the other peoples and nations that form the Spanish State. The Generalitat de Catalunya is the self governing body of Catalonia and its Department of Education controls the public administration of education. The handing over of the educational services from the State to the Generalitat, together with the responsibility for resource allocation, began in January 1981 and ended in 1985 with the transfer of the University education.

Throughout its history Catalonia has shown great concern for education, and has always been open to leading European pedagogic thought. Catalonia's educational policies have been at the forefront of educational planning, working towards the improvement in both the conditions and the standards of its educational system. With information technology being a major focus of innovation in education, the Department of Education set up in 1986 the Programa d'Informatica Educativa (PIE) or Information Technology in Education Programme, in order to provide the education system with computing equipment and a wide variety of services to foster the integration of information technology in primary, secondary and vocational schools, and also into special needs schools and resource centres. The Programa d'Informatica Educativa provides in-service teacher training, pedagogical and technical support to schools, information and documentation services, telematic services, fosters curriculum development projects, and is active both in pedagogical research using information technology and in development of software materials.

Catalonia was proud to welcome participants from over 25 countries worldwide to the IFIP WG 3.1 working conference in Barcelona.

IFIP WG 3.1
Working Conference
Barcelona, 17th - 21st October 1994

Integrating information technology into education

Words of Welcome given by
Sra. M. Angeles Gonzalez i Estremad
Directora General d'Ordenacio Educativa del Departament d'Ensenyament de la Generalitat de Catalunya

Both on behalf of the Conseller d'Ensenyement and myself, I would like to welcome you to Catalonia, and its capital, Barcelona. Participating in this IFIP Working Group 3.1 Working Conference "Integrating Information Technology into Education" are representatives from more than 25 countries world-wide.

The extraordinary development of information technology over the last ten years has generated overwhelming interest and activity, an example of which is this Working Conference. It is not possible to comment upon here, in these brief words of welcome, the diverse and profound impact that IT has had on society which is so apparent in the industry and service sectors, in applied and theoretical research, and in the worlds of intellectual work and entertainment.

Education today cannot be distanced from this situation nor the problems and challenges that the future holds for both students and the welfare of society. There is widespread interest in the ways in which IT can be integrated into education in a beneficial manner, and this can be seen in many ways: the formulation of specific IT in education policies in many countries of the world; in the allocation of important economic resources and efforts in human investment; in the many educational innovative initiatives led by the professionalism of teachers; and also in the increasing number of international opportunities to exchange information, knowledge and experience.

We believe that Catalonia, because of its educational achievements over the last few years, is in an excellent position to hold an international conference upon the integration of information technology into education.

This is because we have highly trained and motivated people with a wide scope of experience. In Wednesday's presentation you will have the opportunity to know about our aims and achievements in this field.

I take this opportunity to point out that the Catalan government encourages and supports this activity. Our government aims to give all pupils in the education system within the obligatory age range (6 - 16 years) the opportunity to acquire both knowledge and practical skills for information processing and technology by "living" this technology at school.

The knowledge and skills which they develop will widen their personal prospects, will facilitate day to day learning, will aid further and advanced studies, and will improve their position when joining the labour market, resulting in economic and industrial development and in greater creativity and job satisfaction. Additionally they will develop their own personal vision of information technology, which is a very important factor in its control and its use in serving individuals and society.

The implementation of these aims will help to build a general technological culture and will improve the equality of opportunities for all. These orientations clearly coincide with the directives of prestigious international bodies such as the Organization for Economic Development and Cooperation (OCDE), the European Commission and IFIP itself.

We appreciate the confidence that a prestigious international organization such as IFIP places in Catalonia and her people in order to host one of its most characteristic and profitable activities: to receive and unite experts from all corners of the world to promote the dissemination of information and the creation of knowledge in such an important field, both for education itself and for the contemporary society.

We are certain of the personal value of having the opportunity to know about a wide range of situations, approaches, and experiences, as well as of the importance of the contributions that will emerge from this Working Conference through joint collaboration in this international environment of highly qualified and motivated professionals.

Ladies and gentlemen, I welcome you once more to our country, and wish you an enjoyable stay and a successful conference, and I hope that the memories of your time here will persuade you to revisit us soon.

I declare the opening of the IFIP Working Group 3.1 Working Conference "Integrating Information Technology into Education". Thank you very much.

1

New technologies: integration into education

Bernard Cornu

Institut Universitaire de Formation des Maîtres (IUFM)
Grenoble, France

ABSTRACT

New technologies are penetrating education. But new technologies are usually simply added to other topics in schools, not really integrated. New technologies are not widely used in schools as one would have expected some years ago; the generalization of new technologies has become one of the main problems, and integration is now a necessary step. Integration can be defined as "combining parts in a whole". In this paper, I examine the question of integration in several aspects: hardware and software integration, integration into disciplines, integration in teaching and learning. In particular, I discuss the evolution of the profession of a teacher towards an "integrated" profession, that is with many different competencies which are not only juxtaposed, but which are interact, and I draw some conclusions for teacher education. Education now needs to design integrated environments; this is a huge task for educators. This paper poses a number of questions about integrating new technologies into education, in order to evolve from juxtaposition to integration. Only when new technologies are integrated, will their use become natural, easy, and will they have a wide effect on teaching and on learning.

Keywords: integration, organisation, policy, social issues, teacher education

INTRODUCTION

In most countries, education has been penetrated by computers and informatics. Schools have computers, a large number of teachers use computers and new technologies, and textbooks often have some parts devoted to new technologies. Many experiments have been done, many journals are published about new technologies in education, many conferences are held about that topic. The development of new technologies in education, however, is not as wide spread as one would have thought ten years ago, and a lot of progress has still to be made. Computers are more and more sophisticated, but also cheaper and cheaper, and more easy to use and to carry. Software is increasingly user-friendly, and adaptable to educational uses. Teachers are offered pre- and in-service training. Meanwhile in society, new technologies are developing quickly, and are used in many domains in everyday life. So, why are new technologies not as widely used in education as one would hope?

There are two major reasons : generalization and integration.

Generalization: Much effort has been expended in order to develop hardware and software adapted to education. A lot of very interesting experiments and research have been carried out about computers in education. But computers in schools are used only by some teachers, the most enthusiastic ones, who have to spend hours and hours, nights and weekends, in order to incorporate computers in their teaching. We now need all teachers and all pupils to use computers in teaching and learning. We need to generalize what is now done by only some teachers and some pupils. Of course, this raises a number of questions. How can the use of new technologies in education be generalized? What are the necessary conditions? Are there helping factors? Certainly, some answers can be provided by appropriate equipment, and by appropriate teacher education; but they are not enough.

Integration: New technologies are most of the time just added to other topics in schools. Informatics courses are added to the curricula; a computer room is added to the other rooms in schools; time for new technologies is added to the pupils time-table; a chapter about the use of new technologies is added to the school books; computer activities are added to the other activities in the classroom. But they are time-consuming, and the teacher must then squeeze the rest of the curriculum into the remaining time. We now need a clear integration of new technologies in school, not an addition: integration in subjects, integration in teaching, integration in learning, integration in the school, integration in the profession of the teacher.

The question of integration is wide open; the aim of this paper is to focus on this aspect of new technologies in education, and to raise some questions.

Certainly, there is a link between generalization and integration. If only integration was achieved, generalization would follow; and conversely, generalization of the use of new technologies by teachers and pupils would impose integration.

THE GROWTH OF INTEGRATION

What does the word "integration" mean? The dictionary says: "combinc parts in a whole". We will then have to study what the parts are, and what the whole could be, and which whole we would wish for. The dictionary also says: "incorporation of new elements in a system", or "co-ordination of the activities of several organs, for a harmonious working". There are many examples of integration. One speaks about the "integrated circuit", "integrated kitchen", "integrated processing" and "integrated management". Integration is also often used in the sense of the opposite of segregation, bringing everybody into equal membership in society, disregarding race or religion. In France, some years ago, we had a minister of integration.

Computers and software arc more and more integrated within themselves: the evolution of computers is towards more integrated material. Computers now integrate many different facilities, and you can find notebooks with an integrated screen, disc driver, printer and modem. The size and compactness of computers allow for more integrated hardware. But what also makes computers increasingly integrated is both the software and the system. You can now use simultaneously different software and facilities. Integrated software provides word processing, spreadsheet, data files, which you can use at once and which are linked to each other. Many specialized applications are now facilitated by integrated software devoted to that particular field, for example computer assisted design, computer and music. CD-ROM, videodiscs, have provided a new step in integration, giving the possibility of mixing words, numbers, pictures, sounds and movies, and processing them as objects in a similar way. Thus you can cut and paste a part of a text, a part of a movie, a part of a picture, a part of a piece of music. You can incorporate music in a text, in a picture, you can incorporate a video in a text. Texts, sounds, pictures, are considered as similar objects, to which you can apply the same operations. In that sense, they are integrated: as parts, they have been combined into a whole.

Multimedia is an example of integration. It is not only the possibility of using different media, but the possibility of linking them, making them bound up with each other, that is significant.

Information and communication technologies are now linked and integrated. Computer, telephone, fax and television. Allow us to imagine in which direction integration will develop. Many tools which seem independent are becoming linked, fields which were originally distinct are now closely overlapping, such as informatics and audio-video. Many of us have at home a radio, a television, a tape recorder, a video, a compact disc system, a telephone, perhaps a fax. All these materials are generally distinct. Increasingly, however, we connect them together through a huge quantity of wires. This is the first step of integration. And probably all these objects will one day be integrated into a unique one. The old black vinyl records for music are disappearing, giving place to compact discs. Books, which take so much space, will be in compact disc format and linked to our computers. Electronic dictionaries and encyclopaedias are already providing us with good examples.

But "integration in themselves" does not mean that new technologies are integrated in society. That is another step yet to be taken.

New technologies will be integrated in society when they will no longer be a supplementary tool, added to what existed before, but when will take their place and become natural and "invisible". It is already the case for some technologies such as telephone, television and pocket calculators. Fax and "Minitel" are starting to be integrated. Computers are becoming integrated for some partial or specialized uses such as automatic bank cash, machines delivering train tickets, and some household apparatus, they but are certainly not totally integrated in society. Nevertheless, evolution is clearly moving towards an integration of new technologies in society.

But what about education? The integration of new technologies in society should imply their integration in education. I will try to describe different aspects of this integration.

INTEGRATION IN EDUCATION

New technologies are integrating into disciplines and more disciplines are being influenced by new technologies in an integrated way. Most significantly, some fundamental concepts are evolving and new concepts are appearing under the influence of new technologies. These new technologies and their own concepts are now involved in many disciplines. Researchers use new technologies in almost all subjects, and this changes the nature and the methods of the work in each subject, for instance, in mathematics the concept of proof, the concept of function and the role of computations, have changed with the use of new technologies. Indeed in many subjects, new possibilities for statistics, for data processing, for simulation and experimentation, have changed the concepts and the methods through their integration in the subject. The frontiers between disciplines have also changed under the effect of new technologies, making some of them closer.

What is true for disciplines at the research or professional level is not totally true at the education level so that disciplines as they are taught do not always use new technologies. The transposition from scientific knowledge to "knowledge to be taught" has not yet taken into account new technologies, and there is a need for reflection about the integration of new technologies in "disciplines at school" and about "integrated curricula". An important question for each of our countries is "how to design integrated curricula", that is, curricula that actually integrates new technologies?

Pedagogy
New technologies provide tools for teaching. Currently, these tools are generally used as supplementary tools besides the usual ones. Integration should eventually lead to "integrated teaching". This requires considerable reflection, not only about the way tools can be integrated, but also how "integrated pedagogy" could be designed. An integrated pedagogy uses new technologies as a fundamental component. It is not enough to integrate new technologies into pedagogy: one must design afresh an integrated pedagogy, new integrated pedagogical methods and new integrated pedagogical tools.

The teacher
The profession of a teacher should be an "integrated" one. A teacher needs many different competencies. First teachers must master the subject they have to teach. But this is not enough. They must also be able to step

beyond the immediate confines of the subject and observe it from outside, know about its history and evolution, and about its applications in society. Teachers must master teaching and learning, and the didactical aspects of the transmission of knowledge. Teachers must know about pupils, their psychology, their behaviour in a group. Teachers must know about the educational system, its role in society, and the ethical aspects of the profession. All these components must not just be added, juxtaposed one to another. There are strong interactions between these components, and the teacher must integrate them and thus combine them in a whole.

It is similar with new technologies. Integration means the possible intervention in the different components of the teacher's role, in the discipline, in the transmission of knowledge, in pupil and class management. This needs both appropriate tools and appropriate scenarii. Designing them is a substantial issue. When designing new software, new tools, new strategies, all these aspects should be taken into account. New technologies themselves can help this integration, because they facilitate communication, linking and interactions.

The changes in society, among pupils' perceptions, and the evolution of new technologies, are leading to a new profession for teachers. Will this new profession be more integrated? What are "integrated teachers"? Their job is not only teaching twenty hours per week in classes. They spend as much time working outside the class. Certainly they will need an office in the school, well equipped with new technologies. They will work together with colleagues, normally in groups. They will use new technologies in the classroom. They will not only be delivering knowledge, but acting also as counsellors, advisors, organizers, leaders and managers. They will collaborate in the design, elaboration and production of tools for teaching. Thus they will be genuine intellectuals, specialists in teaching and learning, true professionals. As good professionals, they will have access to the best and most efficient tools, and they will be prepared to use these tools and be able to make informed choices of the most appropriate ones for them.

New technologies imply new competencies for the teacher, but also make the traditional competencies more necessary, more linked to one another, and more integrated. An important area of study is how new technologies will be integrated into the profession of teachers.

Learning
From the learner and learning point of view, integration leads to another series of questions. The different components of learning should not be separated, but linked and integrated. How can new technologies help for

an integrated learning?   How can new technologies help to take into account the learners' approach?   How can new technologies help in individualizing teaching and learning?

For example, evaluation is generally not integrated in learning; rather it is often considered as a control after the learning. Everybody knows that evaluation could contribute in learning, and could be a part of learning. Do new technologies help to integrate evaluation in learning ?

The meeting of "neurosciences" and "cognisciences" with new technologies provides an opportunity for integrating new technologies into teaching and learning.   It can allow for a modelling of knowledge and modelling of the learning, and then implementing, experimenting and simulating them, leading to new tools and strategies for teaching and learning. Work in this field has already begun.

Integrated environments
The integration of new technologies requires newly designed integrated environments.   In schools, hardware should not only be placed in specialized rooms, but be integrated in the usual rooms. In such a room, new technology tools should be available easily, without particular preparation; they should be "invisible", in the sense that a telephone is "invisible" in a house, so that you don't notice it and you use it without specially thinking about it. Such an "integrated classroom" should contain basic tools such as overhead projector, video, computer, and perhaps be linked to a network. A school should also integrate new technologies, and reflection about the reality of an "integrated school" is an interesting debate. There do exist some examples in different countries. The tools and resources one uses in teaching need also to use new technologies. In textbooks, for instance, new technologies must not be in an additional chapter, but must be integrated throughout the book. Multimedia provides a new style of "new technologies textbooks", mixing a wide variety of data and resources. School organization in general, and details such as the pupils' timetable, need to be thoroughly integrated.

Designing such integrated environments is, however, a huge task for educators.

IMPLICATIONS FOR TEACHER TRAINING

The evolution of education, the changes in the profession of a teacher, the need for generalizing and integrating new technologies, make teacher training a pivotal point. In teacher education and training, integration is again an important issue.   Teacher education must incorporate both knowledge acquisition, and professional and pedagogical competencies. These are not two separate fields, but need to interact.   Knowledge and pedagogy must be integrated in order to build the professional ability of a future teacher.   Teacher education is based both on theoretical and academic training, provided in appropriate institutions or universities, and on practical training, through schools and classes.   Again, these two components of teacher education are not independent, but must interact, must be fed one by the other.   It is again a matter of integration.

Future teachers do not teach the way we tell them to; they reproduce the way they are taught.   So in teacher education, the methods and the pedagogical strategies which are used are at least as important as the content.   This is particularly true for new technologies.   In teacher education, the first need is not courses about the use of new technologies, rather the first need is to actually use new technologies in the training. Future teachers need to learn something new through new technologies, to have the feeling that new technologies brought them something they would not have acquired otherwise.   It is therefore absolutely necessary to integrate new technologies into teacher education. This is a difficult question for teacher education institutions, as well for pre-service and for in-service training.

CONCLUSION

I have argued that the question of integration should be raised at many different levels: disciplines, teaching, teachers, learning, education environments and teacher training. The integration of new technologies in education needs some reflection about the educational system of tomorrow, and the nature of the school of tomorrow. In every country, in every school, we need development projects, exploring the reality of integrating new technologies. It should be a matter of policy, at the national level, as well as at the school level. Organizations such as IFIP could provide an arena for reflection, orientations and guides for good practice. Curricula, teacher training, school organization, school policies, are interrelated topics of concern.

A world wide perspective, dealing with the different questions not separately, but with a coherent and global approach, will help to improve and define an "integrated educational system". Of course, an educational system cannot be examined without reference to the other systems in society.

I believe that we have to make a move from juxtaposition to integration. New technologies were added into the system. They were then incorporated within the system but as a distinct part, reserved for specialists. The next step is to be inside, fully integrated and institutionalized.

But integration must enable us to change and evolve. We must not integrate new technologies in our old habits, integrate our old strategies in new technologies, but we must change our minds, change the system, change the arena of problems. We must have a global and dynamic reflection about integrated tools and methods.

When new technologies are integrated into education, they will no longer be "visible". You do not notice what is integrated, you use it without thinking, it becomes as natural as a telephone or a watch.

All the papers in these conference proceedings contribute by providing examples of integration and answers to some of the related questions. In this way this conference is part of the reflection and debate which I consider essential for successful integration.

Bernard Cornu is a Mathematician at Grenoble University. He studied the influence of computers and informatics on mathematics and its teaching, and also worked in didactics of mathematics. He was the Director of the Institute of Research on Mathematics Teaching (IREM) of Grenoble, and then the head at the in-service teacher training office for the Academy of Grenoble. He contributed to the French reform of initial teacher education, leading to the creation of the University Institutes for Teacher Education (IUFM), in which all primary and secondary teachers, in all disciplines, are trained for two years. Bernard Comu is the Director of the IUFM of Grenoble, and was until 1994 the Chairman of the 29 IUFMs in France. He is the vice-chair of IFIP Working Group, 3.1 and Chairman of this Conference.

2

The computer: ally or alien?

Charles Duchâteau

Faculté Universitaires N-D. de la Paix
Namur, Belgium

ABSTRACT

Nearly all western educational systems are in a deep crisis; divided organisation of the school and standardised views of the teachers tasks, dating from a time when the missions of the school institution were rather different, are obviously one of the causes of this observation. It is in this context that we have been trying for many years to associate the computer (and information and communication technologies) with teaching practices. This paper aims to show why and how these tools and the related approaches bring an innovative and disturbing pedagogical model; it will also explain why and how these new modern tools could contribute to the necessary changes of educational practices.

Keywords: attitudes, classroom practice, integration, interdisciplinary, teaching methods

INTRODUCTION

Today school: long live Socrates
For about thirty years, western societies have been asking schools to welcome and to train all children between six and sixteen (or eighteen) years old. As Lafontaine states [1]:

"In a period of twenty years, the nature and the extent of contributions and functions expected from school have deeply changed. From a purely quantitative point of view, school was confronted with huge number of pupils. In Belgium, between 1950 and 1965, the total number of students at secondary schools increased by 75%, and by 29.5% from 1970 to 1982. In France, the number of students at secondary schools increased fourfold between 1959 and 1985 ... we can easily understand that school was faced with the need for an urgent solution to this massive increase in the number of *different* students, different from the usual school clients."

Hence, the number of pupils as well as the number of teachers is now greater than ever before. Yet, neither teacher training nor school organisation has taken into account these transformations.

Our teaching model remains - at best - the Socratic one with masters helping chosen disciples to find their own way towards wisdom and to build their knowledge through a particular and adapted dialogue. This experience of learning, thanks to the contact with a master, is a very intense but an unusual one in the average pupil's school, and I do not aim to call this experience into question in this paper.

The absurdity, however, lies in the belief that this method can be, or even is, generalisable and that training can be organised just as if all the "teachers" were always and in any circumstances real "masters". The present organisation of educational institutions is based on a dull and absurd equation: "If one Socrates is enough for training and tutoring twenty disciples, how many Socrates do we need for the training of two million pupils?"

In the economic world, the need for mass production leads to giving up the old craftsmen model and to developing industrialisation. Today school is faced with the same need for an "industrial revolution": the task is no longer to train a small fringe of first-rate pupils but to train at best all children from 6 to 18 years old. Yet, the model of the teacher still remains the one of a craft-teacher who is able to teach in a classroom, to prepare for this teaching, and to cope with psychological and relational difficulties. Even if we tend to prefer the good old days of craft-teaching methods, school has today to undergo its own industrial revolution; to knock down all the barriers between classes, levels, subjects and to enhance the vision of what a teacher is and what a teacher does. I believe that a shift from a stereotyped and uniform view of teacher's activity during his whole career to a broader and more evolving view of this profession is essential. Unfortunately, this observation is not really new. Hyer wrote in 1974 [2]:

"Within the next ten to fifteen years, major changes will have to be made in the training and retraining of teachers. More emphasis will need to be given to individualising instruction, operating as a member of a team, assessing pupil achievement and diagnosing learning difficulties, providing a working knowledge of technology and of man-machine relationships, and the selecting, modifying and/or producing of instructional materials."

Of course, the aim is not to transform classes into assembly lines with a mass-production of similar pupils and to introduce Taylorism into the school. To talk about industrialisation in this context means, on the contrary, to change the role and the function of teachers: sometimes they will teach in the classroom, or rather, they will assure that learning takes place; during other periods of the week, of the year, or of their professional life, they will make tools and develop environments for teaching and learning; lonely work will give way to team work and the sharing of common tools.

Nevertheless, as pointed out by Vitale [3]:

"In spite of innovators' repetitive efforts and despite partial and limited experiments carried out in special or experimental schools, today school is dominated by the presence of notional knowledge already made up (only known by teachers at the beginning but slowly passed to students in a hard process of knowledge socialisation) and by the cutting-up of human knowledge into subjects whose frontiers are jealously watched over by experts."

As Hannafin also mentioned [4]:

"A typical classroom, both past and present, may be described, with notable exceptions, as having one teacher who directs all activities and presents knowledge in discrete chunks to be passively ingested by students to be recalled later on in a test."

Finally, there are no simple solutions or cures to this school crisis; we have to be wary and careful about specialists making recommendations and creating fantasies. Yet, we have to go on analysing the problems and we have to find and test out solutions, even if these solutions are incomplete: the survival of school, hence of our society and our culture depends on it.

The computer
A group of specialists in education have considered for some time the computer with its range of uses as a possible means to diversify, develop

and even to improve the pedagogical relation in its teaching component as well as in its learning aspect.   From obsolete teaching machines to the prophecies on the revolution linked to the generalised distribution of pocket computers, the new technologies have raised hopes.   Even if achievements have probably not always lived up to expectations and to the quantity of speeches, it is important to continue considering possible ones.

In this paper I will focus on some potentialities of these new information and communication technologies by stating in what way they operate and how they are likely to intervene in pedagogy or in the routine of the trio teacher - student - subject.   So I do not wish to point out all the applications and possible uses of the computer, but only to highlight some of their possible effects on teaching, learning, the organisation of the school and the perception of the role of the teacher.

THE COMPUTER: A TUTOR, A TOOL FOR THE TEACHER OR A TEMPTATION FOR THE STUDENT?

I would like here to examine three possible meanings of the acronym CAL by focusing on the position of the computer in the classical triplet subject - teacher - student.

CEL - Computer-Ensured Learning: the computer-tutor

Today this vision, fortunately out of date, of a first possible role of the computer, has perhaps been the most widespread illusion: the role of a tutorial.   It involves the type of use which many people deny having done but which, however, has deeply influenced the beginning of the use of the computer for pedagogical purposes.   It is the illusion of the computer tutor, that of computer-based learning and that where the software is trying to mimic as far as possible the actions and the reactions of a good teacher.

The student is placed in front of the machine-tutor and after the dialogue which is going to occur and which will be, of course, individualised, the student will learn and master, for example, the concept of the derivative or the workings of the combustion engine.

The vision of the learning conveyed by this type of approach was originated from the Skinnerian or Crowderian trends which had already widely influenced programmed teaching. The contents to be mastered are dissected and split up in small cells which will be successively proposed to the learner. According to the answers given, the learner is dragged along a more or less individualised learning track. At the end of the dialogue between the learner and his tutor, the concept is mastered and the subject is digested.   One may imagine the huge difficulty of this prior dissection of the subject to be taught which requires both a great expertise in the contents, but also a deep knowledge of the difficulties encountered during learning. It requires a complete vision of the subject to be taught and its organisation.

When we look at them from the design or analysis and programming point of view, the creation of the information technology tools is characterised by the fact that the task to computerise must be completely unfolded beforehand. No vagueness, nothing implicit is allowed, even in educational software tutorials. The designer must therefore have considered beforehand the dialogue which is going to constitute the learning but only has access to only one of the two protagonists. It is only possible to write the role of the computer with all the subsequent responses to all the predictable reactions of the learner. The problem is therefore no longer to teach but to make the machine teach; this *machine tutor* is particularly limited.   Taking the main characteristic of this performer into account constitutes the second difficulty.   Beyond the extreme and total dissection of the contents to be taught, there remains a detail which is at the heart of informatics; the great difficulty is that the computer is a formalist processor of information. In the word informatics, what is to be read is not information but form.   The real objective of informatics is to detect the meaning under the form, to enclose the meaning in the signs and the semantics in the syntax. The domain which is the most out of reach without a doubt, is the one of the language and even more that of the dialogue.   Therefore, a great deal of time has been lost and much energy spent on a problem which is, in the end, technical and marginal, that is the analysis of responses coming from the learner.

There is an additional illustration of the false importance granted to this type of use.   Many proposals of evaluation of the CAL products are made by observing and judging what the computer does and not what the teacher or learner does.   If one wants to judge the degree of integration of a software tool to the strategies of the teacher, it is the teacher who is to be looked at and if one wants to assess the qualities of the learning taking place, the student and not the screen, is to be watched!   Yet as Crossley and Green state [5]:

"In many educational software, the student is considered as a pitcher which has to be filled up. Some educational software designers anticipate a passive student, taking on knowledge and never tempted to switch off the computer. With their software, it is the computer which has fun."

It is possible that, in medium term and for a particularly limited universe, the progress of information technology will perhaps enable us to tackle this central problem of the CAL tutorial, that is the management of a pre-recorded dialogue. Attempts must no doubt follow but with a clear measure of the challenge. These procedures, however, which aim at changing the computer into a private tutor should no longer continue to constitute the tree which hides the forest from possible pedagogical applications : there is so much to do with a computer than to try to make it mimic a teacher. As pointed by Barchechath and Pouts-Lajus in Crossley and Green [5]:

> "it is wrong as for designing educational software ... to believe that the only way to reproduce an interpersonal communication between an interactive software designer and a user imposes to the designer that the software simulates a person (and for instance, the designer himself). Yet, the sole observation of educational software users interacting with informatics systems, could have presented a precious indication as far as the validity of the similarity necessity is concerned: none of those users claims that he thinks or that he wants the computer to behave like a human being."

CAT- Computer-Assisted Teacher: the computer set of tools

This is no doubt the domain where we could harvest a wide variety of products and ways of uses. They all have in common the fact that the teacher again assumes a role and that the computer, equipped with adequate software packages, returns the role which it should not have left, that is a tool or rather a wealth of tools. It is of course arbitrary and delicate to classify the products under the category of tools for the teacher or those for the student. Some of them can of course, be fortunately used by either of them. Without being exhaustive, I would like to point out a few typical and well known examples of this type of uses.

The computer "super-board" involves activities which the teacher decides on, prepares and carries out in the context of a traditional lesson. In front of the class, near the board or the overhead projector, the teacher uses the facilities of a spreadsheet or graphical software to illustrate a concept, to shorten repetitive and tedious calculations and to deal with the data of an experiment.

The dull drill-and-practice programmes are those which, for obvious reasons, restrict the dialogue between the student and the computer to an exchange such as 'delete as appropriate'. The dream of the tutor-computer has disappeared and handed over to a limited and patient auxiliary, just good enough to propose statements and to indicate the correction of the stereotyped answer provided in the range of offered propositions. The initiative of the student is extremely reduced especially because of the form of the possible answers.

Everyone knows that a part of teachers' work consists of activities not requiring a great deal of grey matter, for example the acquisition test of the students' reflexes, propositions of practice exercises and their correction and checking of factual knowledge. These are typical activities where the computer equipped with little sophisticated software is irreplaceable. It is better to confine it to the role of a repetitive machine than to that of a private tutor. It does not matter here that the tools used are dull; they come to assist teachers with dull activities with which they are grappling. Then relieved with repetitive and tedious aspects of their task, teachers will perhaps find some time for remedial work which constitutes a major part of the task and for which the teacher is irreplaceable.

Even if teachers are absent, as long as learners are kept busy with the software exercise, they are the ones who decide, organise and manage the access to this type of product; in a modest way, teachers then become the managers of this particular form of learning for learners.

Bureaucratic software packages, however, are of a different type. I believe that for teachers to be assisted by a computer in the classroom, they must have perceived the advantages of these tools. Using a computer for personal purposes, material preparations, and the management of reports; these are decisive in arousing a more pedagogical curiosity. The most efficient products in this context are obviously the bureaucratic packages. At the same time, once these tools have been mastered, one can then think about their possible uses in the classroom, consider the pedagogical diversion which these packages promise and to experiment with them. These packages are, by nature, open to a wide range of uses; here too, it is the imagination and the creativity of the teacher which constitute the main motor for developing these activities.

This list of types is far from being complete; I have stressed that it is the use which is often more decisive than the characteristics of the products. Tutorials can be used by the teacher as a remedial tool; a management package as a super-board.

These uses have in common the central role of the teacher; it is the one teacher who decides where, when, how and which tools to exploit and to be used by the students. If the pitfalls of CAL mainly appeared at the design level of the products, the difficulties of this other side of CAL appear at the level of changes in the roles and attitudes of the teachers. It is obviously impossible for a trainer to integrate these new tools into practices and make them implicit without realising it. Thus it is far less the products than the uses made of them which need to be evaluated. The question to ask is not "what does the computer do?" but rather "what does the teacher do?".

CAL - Computer-Assisted Learner: the computer-world to explore
This is no doubt the rarest type of use, but also the most promising. If I were to qualify it in a few words, I would say that as with video games, learners are found engrossed in a little world which they are going to explore gradually. The actions posed are going to cause modifications of the environment and force the students to construct hypotheses which will then be tested thanks to the interventions on their part. Little by little, with the help of their companions or the teacher, they are going to reconstruct in their heads the world on which they act and the rules which govern it; thus they are going to build themselves new knowledge. There still exist a few of these tools which allow the exploration and discovery of a particular subject. I will only quote as an example Logo or Cabri Géomètre.

If teachers still remain the managers of the learning environment, they may often also become the accomplice of the learner and the co-explorer during the discovery of the micro-world offered by this type of software. The role of the guide, in the touristic sense, should not be overlooked either: if teachers accompany the students in their discoveries, they are going to indicate and highlight the important details, make them aware of the learning taking place by analysing the route they took and show the enlightening analogies. This role of putting things into perspective has been stressed several times as crucial for allowing the transfer of acquired skills.

The real problem is once more not the technical mastery of these tools, but their integration in a global strategy and the teachers' perception of the change in the role which is required of them at certain times.

THE COMPUTER-LEARNING OPPORTUNITIES

I am tempted to state that in the triangle subject - teacher- learner, it is the vertex subject that disappears. As it is, the main goal is no longer to learn and master a traditional subject, but only to learn. The computer is the pretext or excuse for learning activities that will bring a cognitive surplus into the learner's brain.

The first and most prestigious representative of a learning opportunity through using the computer is undoubtedly Logo. It is also the more widely used and analysed example. I would like to insist on the fact that through the Logo micro-world, learners will explore and experience algorithmic thinking. Even if it is done in a soft and unconscious way, the learner who explores the Logo universe is walking in the heart of algorithmics. Working with Logo, the problem is not to draw a circle, a house or a ship, but to have these pictures drawn by the turtle-performer, by giving it the suitable directions. The learners do not draw anything, they have it drawn by the turtle, so they program. Through Logo, you go from the usual world of 'doing' to the new and fascinating world of 'having it done by'; so you are at the heart of algorithmic thinking.

We all know that the main goal of Logo environment is not to learn geometry, even if the tool can be diverted for that learning, nor, despite the previous assertion, to learn programming. Logo activities are free; the aim is not to teach a traditional subject but to train learners to learn and to think about their learning.

Using some software as tools also brings opportunities for training some methodological competencies which would allow the pupil to do at school what is usually supposed to be done at home.

"Many skills have to be learnt during schooling: to be able to look for information, to process it, to communicate it, to interpret it, to be self-taught in individual or collective research activities, using recent equipment. By using intelligently a word processor, a data base or simulation, you can practice those skills." [6]

In this case too, these tools are pretexts for developing organisational, planning, and collaboration competencies.

We should also remember the field of pedagogical robotics. Giving orders to such a performing device (robot, automatic machine) even if it is very close to algorithmic thinking, has no importance by itself. What matters is not to have the task done by the robot, but rather the approach and processes leading to this result - questions, hypotheses, attempts, testing, error recognition and team working.

Although from the learner's perspective these opportunities may seem to be free, the disappearance of a precise and classical teaching about a subject may be quite disturbing for the teacher, who consequently requires a particular training to this kind of pedagogical way. This implies that, more than ever, research, experiences and exchanges have to be emphasised and communicated. We know that to put pupils in these rich learning environments is not enough for them automatically to acquire new competencies and knowledge. Teachers have to be specially trained in these new roles to become guides and accomplices of the learners.

The subject of this working conference is Integrating Information Technology into Education and not Integrating Information Technology into Existing Subjects. I believe that the computer provides a wonderful opportunity for inter- or multi-disciplinary activities, which are not limited by the existing subject frontiers [3].

The computer-mirror

In order to bring and to integrate the computer into pedagogical practice, we need some prior reflection about these practices; we could even say that the computer will give us a reflection of these practices, since it can be considered here as the mirror of actual pedagogical behaviour.

This computer-mirror asks many questions about learning processes and forces specialists to model and to formalise these processes; it also questions teachers about the ways they use it through teaching practices and it compels didactic experts to have a new look on their own subject. All these remarks were already pointed out by Bertrand [7]:

> "but it is as essential to notice that informatics highly contributes - in the learning field - to bring up to date buried problems: what is *to learn*?; how to learn complex mathematical notions more efficiently ? ... In that sense the trick of informatics fulfils its function as it forces man to wonder about himself"

Also in Jaffard [8]:

> "And finally, the presence of the intruder (computer or IT) leads the teacher to wonder about his daily pedagogical practices ... to call some of them into question, to invent new pedagogical situations".

And in Beaufils [9]:

> "In some cases, indeed, the use of computer underlined insufficiencies in the knowledge or the know-how of those subjects."

From my own experience, when working with a team of teachers, creating some educational software or thinking about the ways of how to use an existing one, you always, first of all, have to cope with present, current and usual practices. You have to be aware of your own teaching practices before deciding why and how to use the computer. But this is the usual consequence of informatisation; you have to depict and to become conscious of the present situation before using the computer. This is the mirror-effect of the computer in school as well as in business; the computer obliges us to drive out all tacit and vague knowledge. It is a mirror reflecting what is really the reality.

The computer-lever
Insofar as the computer is a learning opportunity for the pupil, it can represent a chance for the teacher to change role and strategies. I stressed before the roles of guide, of co-explorer and of accomplice. We could add many other roles to this list, all of them sharing the same goal: a shift from loud-speaker behaviour or knowledge funnel role, to learning environment and learning scenario producer; to resource-person, to a cartographer drawing maps for exploring knowledge lands, these lands being not enclosed in the artificial barriers of subjects. Meirieu states [10]:

> "Relieved from pure informative tasks, he (the teacher) could dedicate himself to the processing of these information : he would guide the students in the multitude of diversified documents, would help him to make relevant choices and to make the most of them through efficient working, would not hesitate, if necessary, to send him back to resources from cultural, economic or social field. Because he should not fear to lose his power, convinced that he would barter the distributor role for the mediator one, to become guarantor of assimilation and no more spectator of incomprehension."

I do not intend to refer to a wait-and-see teacher; we all know that giving a bathtub to a student is not enough for him to discover naturally Archimede's principle. We also know that sending him into a library in order to establish a comparison between the US and Japan economies will not automatically lead to an answer, nor to a formative and organised behaviour. We do not have to confuse putting the stress on pupil's activity with dreaming about natural discovery methods. It is not true that the student, on his way to learning, will have sparks of genius which would be the same as those which have marked out the whole humanity evolution. The computer is only a lever which, introduced in teaching methods, can contribute to change viewpoints and roles.

Only an Archimedes could think that a lever and a fulcrum would be sufficient to lift up the world. Whatever the potential qualities of computerised environments, men and women in the educational world will have to lean on this computer-lever to try to modify school, to awaken teachers and to shake pupils.

"If, as I think, programming and other forms of computer experiences will be able to play in the future, a catalyst role in the invention of new relationships between teachers and learners in the classroom, that will be because teachers and learners will have accepted to leave their present roles, too rigid, in order to have a more open one: curiosity, discovery, search for relevance in the building of knowledge. But the presence of computers in school and the learning of computer science won't be enough to change deeply the school landscape. In a rigid school, informatics will be considered as an additional subject, as abstract and irrelevant as others. On the other hand, the failure or the success of a social project of open and interesting school won't be due to the only presence of the computer." [3]

The computer-lever must not become a rubber repair patch computer to be put on school weakness to hide how large and deep are these rifts.

The computer-cement
The computer can also be a wonderful federative tool; I will go even further and suggest that a successful introduction of software tools in the school inevitably requires sharing and exchanges.

The school, which is a public service in our society, has to welcome new sectors, for instance, a sector whose objective would be the creation and experimentation of tools and a proposal of appropriate use of them. Teachers' teams could deal with these new fields, with the help of computer scientists, as we know that team working prevails in this kind of creation.

More widely and within the school, the need for sharing and mastering the same tools can induce exchanges. And if I could be allowed to dream, why couldn't we consider teaching as a collective work, teachers working together, in groups or teams? In a more realistic way, we can assess that the general and open character of some software tools makes easier a multidisciplinary approach.

And finally, outside the school, groups of teachers sharing the same new methods or gathered around the same tools, are emerging. In this case, telematic networks are really useful, and used as sharing opportunities, since their users have to talk together.

New roles for the teacher
This paper could have been entitled Information Technology and Changes in the Role of the Teacher, in the triangle subject - learner - teacher. I hope I have shown that many computer uses, disturbing the usual

equilibrium within a classroom, force the teacher to revise role and to modify attitudes. Some people will talk about the teacher as creator of an interactive learning environment (Oates, quoted in Hannafin [4]), underlining the designing role in a team. Others will consider the teacher as facilitator or manager of information (Berliner quoted in Hannafin [4]); these are only a few examples among many others: coach, guide, organiser, initiator or diagnostician.

Personally, I tend to use similar images, talking about a teacher as a cartographer drawing maps for exploring 'knowledge lands', co-explorer of micro-worlds, accomplice [11,12,13]. Anyway, it is important to notice that those new roles do not constituted a dissolution of the teacher's mission: to be an adult who must teach, raise and educate; to know more than the pupils; to keep defining clearly what they want needs to be learnt and to assess the extent of difficulties met by those who are learning. Mirror and cement, the computer will force teachers to re-define their position and will help them in this task.

CONCLUSIONS

For more than ten years, speeches about the computer-education duo have followed one another. Little by little, more understanding and experiences have emerged. We need to be patient and we need to have a view about these problems that is not too global and comprehensive. In the world of education, it is perhaps better to work with a lowered head, moving small stones around rather than looking at the big, still and unchanging educational mountain. If many of us move these small stones, although we may not move the mountain we may make it tremble. And the computer is such a nice shovel.

REFERENCES

1. Lafontaine, D., Grisay, A. & Orban, M. (1987) *Enseignement et enseignants: mutations et perspectives à l'heure des nouvelles technologies.* Laboratoire de Pédagogie Expérimentale, Université de Liège.

2. Hyer, A. (1974) *Effect on Teacher Role of the Introduction of Educational Technology and Media into American Schools* in The Teacher and Educational Change: A New Role, OECD, Paris, 1974 211-269.

3. Vitale, B. (1990) *L'intégration de l'informatique à la pratique pédagogique.* Volume 1. CRPP, Genève, 1990.

4. Hannafin, R. & Savenye, W. (1993) *Technology in the Classroom: theTeachers's New Role and Resistance to IT.* Educational Technology. June 93, 26-30.

5. Crossley, K. & Green, L. (1990) *Le design des didacticiels.* ACL Editions, Paris.

6. EPI (1992) *Pour une culture générale en informatique à l'école, au collège et au lycée.* Assemblée Générale de l'EPI.

7. Bertrand, I. (1991) *L'informatique pédagogique et l'apprentissage.* Informatique et apprentissages, INRP, Paris 17-27.

8. Jaffard, R. (1990) *L'EAO. Pourquoi? Pour qui? Par qui? Comment?.* Interface, Revue de la SSPCI, **1**, 35-38.

9. Beaufils, D. (1991) *L'informatique en sciences physiques au lycée.* Informatique et apprentissages, INRP, Paris, 1991 47-57.

10. Meirieu, Ph. (1989) *Enseigner, scénario pour un métier nouveau.* Editions ESF, Paris.

11. Duchâteau, C. (1990) *Images pour programmer.* De Boeck-Wersmael, Bruxelles.

12. Duchâteau, C. (1992a) *L'ordinateur et l'école! Un mariage difficile?* Publications du CeFIS, **5.28,** Namur.

13. Duchâtcau, C. (1992b) *Trois visages de l'EAO.* Interface **2/92**, 5-7.

Charles Duchâteau was born in 1946 in Belgium. After studying to be a primary school teacher, he undertook a degree in Mathematics at the Catholic University of Louvain and then a PhD about incomplete information differential games at the University Notre-Dame de la Paix in Namur. It was within the Mathematics Department of this university that he founded the CeFIS - a centre for the training of teacher in informatics. He is now member of the Education and Technology Department. His research field is that of informatics didactics and the meaning of the elements constituting an Information Technology Culture.

3

The roles and needs of mathematics teachers using IT

Bernard R. Hodgson

Département de mathématiques et de statistique
Université Laval, Québec, Canada

ABSTRACT

Teacher education stands as critical among the various issues pertaining to the integration of information technology into secondary mathematics education. Remarkable improvements have indeed taken place recently both with respect to the material aspects of IT (hardware, management, accessibility, maintenance) and the availability of high quality educational software. Successful use of the computer depends essentially on the quality of the teachers and of their pedagogical agenda. Renewed and diversified roles confront teachers in an IT educational environment. Pre-service and in-service teacher education must help all of them to modify their attitudes and develop the new competencies in mathematics, informatics and the didactics essential for them to fulfil their pedagogical mission.

Keywords: information technology, teacher education, integration, pedagogy, attitudes

INTRODUCTION

The last two decades have witnessed a wealth of attempts aiming at integrating information technology (IT) into secondary mathematics education. The advent of microcomputer technology, and subsequent progress both in hardware and software development, have given rise to very high expectations about the impact of informatics on the teaching and learning of mathematics in the secondary school.

The pedagogical potentialities of the computer in mathematics have evolved considerably since the days when it was used strictly in the context of numerically oriented languages. For instance, the availability of the programming language Logo with its underlying philosophy of exploring mathematics in specially designed microworlds, the development of symbolic mathematical systems like Maple or Mathematica capable of carrying out symbolic computations previously considered to exemplify a genuinely "human" ability, or the recent advent of geometry exploration software such as Cabri-géomètre allowing users to manipulate and transform geometrical objects, have drastically modified the perception of the influence computers might have on the teaching and learning process in mathematics. Moreover, the computer capability of handling multiple representations of information - numerical, graphical and symbolic - and of moving from one form of representation to another has opened up the way to unprecedented applications in mathematics education (see [1, p 255-260]).

Nevertheless IT, more precisely microcomputer technology, is probably not as extensively used as an instructional tool as all these developments might suggest, even in mathematics as seen for instance in the recent findings of the IEA Study [2, p 70]. A wide range of explanations have been put forward in connection with this situation: restricted number of computers available in school and their limited power, lack of software of sufficient quality, complexity of user interface, difficulty for the average teacher to use the computer in the typical classroom on a sustained basis [3, p 517]. This paper aims to support the idea that, although all these facts are related to important issues, none are as critical as those pertaining to the teacher's capacity and ability to cope with this new pedagogical environment.

IT IN THE SECONDARY MATHEMATICS CLASSROOM

The successful integration of IT requires a wide range of actions concerning numerous issues (see [2, pp 103-105] or the guideline paper [4, p 22]): general aims of secondary education, expectations of society, curricular reform, material infrastructure (availability and maintenance of hardware, development of software, adequate funding policies), human resources. The public debates that took place in the early years in many countries around the introduction of computers in schools - very often more of a political than of a truly pedagogical nature and held in a nearly feverish climate that could but hinder sensible decision-making (see [5, p 48-55]) - were essentially concerned with material aspects such as: which type of machines should be installed in the schools, and in what quantity? While the numerous difficulties related to the material organization of the IT classroom or to the availability of high quality educational software are certainly crucial, these are not the main impediments to a widespread use of computers in teaching. To a certain extent, remarkable improvements have indeed taken place in recent years in both these respects; for instance a "more than decent" computer environment can now be found in a number of schools in many countries - although on a general scale the actual availability of the computer to the pupil is still rather low. Kaput states that availability "is at best an hour per week per student for mathematics" [3, p 517], while Anderson asserts that a large part of the equipment is becoming obsolete - "50% of the computers used in (American) schools by teachers or students are 8-bit computers" [2, p 21].

Material considerations are not the last word to a successful integration of IT in education, an assertion supported by the main conclusion of the IEA Study:

> "improving education with computers requires more than hardware and software. Students also need to work with skilled people including teachers, parents, co-workers, and friends." [2, p xx]

This paper concentrates on the teachers themselves, their competencies, their attitudes, their perception of their multiple roles. Teacher education (both in a pre-service and an in-service context) is a channel of the utmost importance through which IT can create an impact in the secondary mathematics classroom. This stand-point matches the conclusions of the recent ImpacT Study to the effect that the three main resource-related factors influencing concrete use of IT are:

> "access to computers, the organization of IT in the class, and the teacher's skills and enthusiasm for using IT in the curriculum" [6, p 28];

the teacher's contribution being identified as "the most important" factor [6, p 159]; note should also be made of the remark about the importance of a "sympathetic and positive Head" in developing teacher's confidence in using IT (p 91).

While it is true that modern technology is affecting educational aims in general, the influence is particularly strong in the case of mathematics education, as reported in the Study of the International Commission on Mathematical Instruction (ICMI) [7, p 2, 68]). This study states that mathematics itself and the way in which mathematicians work are being directly modified; the demands, expectations and employment patterns within society with respect to mathematics are changing, the educational goals and structure of entire curricula must be reappraised, and new pedagogical possibilities for teaching and learning mathematics are being opened up. Still Anderson states that, even in the mathematics classes, the use of computers is "decidedly lower than many might have expected it to be" [2, p 68]. It has been claimed about current school mathematics that:

> "our children are taught to do mathematics in ways that are very largely outmoded, with at least 80% of the curriculum time wasted on trying, more or less successfully, to develop fluency in skills of now-limited value." [8], p 9

Extraordinary demands are thus being made on the teacher towards a deeper use of IT in his mathematics classroom.

THE EVOLUTION OF TEACHING

One of the effects of IT integration is to reinforce the "professional" dimension of teaching. While the specificity of teachers' multifaceted tasks has always been such that they should be fully considered as professionals - the "professionals of the pedagogical act" - and trained as such, this professionalism is now stronger than ever in connection with the modified and diversified roles they play in an IT educational environment. It must be stressed however that some of these modifications in teachers' roles are not specific to the IT situation and are of value with or without the computer. For instance, Graf considers that

> "coordination of students' work, conduct of group dynamics, development of problem solving, creativity and reflective thinking by students are examples of roles not always fulfilled in teaching, but not directly related to computer use." [9, p 32]

But the computer environment introduces a renewed perspective inducing a change in both students' attitude towards mathematical content and expectations towards the teacher.

Various roles can be mentioned with respect to the teacher's pedagogical implication in the IT classroom (see for instance [8, pp 27, 88, 97] or [10, p 84]): acting in turn as manager, task-setter,   guide, accompanist, coordinator, explainer, counsellor, leader, resource, and even as a fellow pupil (working on a task alongside the school students). Not any more the sole nor the main source of information, teachers are confronted with their genuine pedagogical mission, that of "facilitators" able to create a context appropriate for a fruitful interaction, via the machine, between pupils and mathematical concepts.

Being no longer those who transmit truth and knowledge to the learner in a more or less dogmatic and unidirectional way, teachers must consequently accept a lessening of the degree of control they exert over the pedagogical activities. They must also accept the resulting shrinkage of their sense of security, in comparison to a traditional management of the classroom - for instance when expected to make sense of a "strange" mathematical object generated on the screen by pupils during an exploration. Teachers must in addition be able to cope with the heterogeneity of pupils' reactions to the various learning situations being offered, the computer making possible a greater diversity of potential ways of reacting than a traditional environment.

The contribution of teachers to the organization of the educational environment has aptly been described by Cornu [8, p 28, 90] and Artigue [11, p 29] as that of a "didactical engineer", in the sense that their work, while being based on a body of scientific knowledge, is concerned with more complex objects than the refined ones of science: their central task is thus to transform theory into usable products. In the case of IT integration, this requires teachers to be able, for instance, to elaborate a sound pedagogical strategy, to design a coherent teaching program, to choose appropriate tools and products to be used. But above all teachers must be in a position to assess the educational value of the technological import so to ensure that pupils are "engaged in a good learning experience" [6, p 76] in keeping with their pedagogical aims and objectives: use of IT in the mathematics classroom is not an end in itself but a way to better understanding by the pupils, as well as a preparation for work force and literate citizens of an information-oriented society [2], p 99].

"The decision to implement a new technology ... is fundamentally an educational decision, to be guided by educational objectives. ... The question to be answered is, 'Will the technology help us do better what we have been trying to do?' " [3, p 548].

In order to make such judgments, teachers must have developed a wide range of new competencies related to various aspects of the teaching and learning of mathematics.

THE COMPETENCIES OF TEACHERS

At least three types of competencies are necessary for teachers in order to adequately fulfil the new roles which confront them in an IT environment; competencies related to mathematics, to informatics and to didactics.

The influence of computers on *mathematics* has aptly been described elsewhere, for instance [8] or [12]. Not only can mathematics be taught differently, but in a very deep sense it is different, a much greater emphasis being placed on numerical and algorithmic processes and on an experimental approach involving exploratory investigations. The evolution of mathematics teaching related to an evolution of mathematics itself, of the way it is being practiced. Secondary mathematics teacher must thus develop a deep epistemological view of the subject, enabling an understanding of its origins, its history, its changing role in the society and its growing applicability.

Teachers must also be comfortable with distinguishing those mathematical skills of decreasing importance, (because they correspond to tasks better accomplished by the machine, from mere arithmetical skills like calculations of square roots to more advanced techniques for algebraic computations or graphical study of functions), from those skills becoming more important. These skills of growing importance include the choice of an appropriate mathematical model, identification of significant parameters, discernment between those tasks which should be ascribed to the computer and those which require insight and human decision, interpretation and critical appraisal of the "answer". Such skills involve conceptual thinking and planning rather than execution of routine calculations. The impact on teachers' responsibilities is outstanding. Dörfler and McLone state:

> "This shift ... will not make school mathematics less demanding. On the contrary, according to every known taxonomy of intellectual activities, the abilities needed to make adequate use of computers (and of calculators) are on a higher level than those for executing calculations." [13, p 79].

The overall effect of such an evolution in mathematics is well-summarized by Mascarello and Winkelmann in the following terms:

> "In total, there can be observed a specific shift in the spectrum of abilities, from precise algorithmic abilities to more complex interpretations, so to speak from calculation to meaning ... In this process the mathematics to be mastered tends to become intellectually more challenging, but technically simpler." [8, p 109].

Proper integration of IT also means that teachers must know about *informatics* in general and that they must be familiar with the computer itself, both in connection with hardware and software considerations. While this does not mean that teachers need to be fully-fledged programmers, it certainly suggests that they should aim at becoming competent enough to "reach beyond the constraints of packaged tool software." [1, p 261]. In fact, teachers should have developed enough autonomy so as not to rely upon external help each time something goes wrong: teachers unable to cope with the basic technical aspects of computers are at great danger of losing credibility among their pupils, often much less inhibited by the "machine". But this level of expertise should not be interpreted as denying the importance of support either by technical staff, peers or IT coordinators in the school. More comments on this can be found in Anderson [2, p 50-52, 104] and Ruiz [4, pp 22, 28, 40]; Anderson remarks that "in most instances, coordinators spend more time helping students ... than they do helping teachers", p.51.

Finally teachers must be proficient with the *didactical theory* underlying their work, so to be able to clearly define their pedagogical projects: what should the pupil learn? can the computer be used fruitfully? and how? Didactics of mathematics has clearly been emerging recently as a branch of science with its own body of knowledge (see [11]). It thus behoves the teacher to gain familiarity with its progress in a range of issues, including the relation between pupil and knowledge, the various obstacles (psychological, didactical, epistemological) that can intervene, the types of errors made by pupils and the information thus revealed on their appropriation of knowledge, and the didactical role of assessment. This didactical background is really crucial, as no machine, as powerful as it can be, can make up for a poor pedagogical program. Cornu affirms that "In no case can the technology replace the pedagogy." [8, p 92]. All this results in a complex situation where the teacher is the one first and foremost responsible for elaborating a coherent teaching and learning context within which integration of IT can become effective.

EDUCATING ALL TEACHERS

While many successful teaching experiments have been reported involving the computer, these are often associated with enthusiastic teacher-pioneers and are very difficult to generalize or even to reproduce. Cornu considers that "Un logiciel ne transporte pas avec lui tous les paramètres didactiques, tout l'environnement pédagogique nécessaire." [12, p 55].

One of the major difficulties in the integration of information technology in secondary classrooms is thus to have all teachers become competent with this new environment, and not only those naturally inclined. This represents an outstanding challenge for both pre-service and in-service teacher education.

Many of the early programs aiming at introducing computers in school rested on a core of highly motivated teachers, so-called "multiplicative agents", expected to pass their knowledge and pedagogical agenda on to their colleagues through a "cascade" effect. But this model has limitations and in many cases the anticipated effect has not materialized, see the remarks by Watson [6], pp 96, 162 and Cornu [12], pp 29-30. What is now needed is a more global approach to the concerns of the teacher in relation to integration of IT in the mathematics classroom. The problem does not concern mainly the necessary familiarity with new technology, but more profoundly the teaching and learning issues being raised. Teachers need to gain the capacity to confront those, to evolve and adapt.

It has been suggested that successful integration of IT in schools requires a new generation of teachers, that Yokochi et al describes as technology oriented teachers [14], p 71. It is clear that we can perceive the emergence of a "new teacher", more open-minded with respect to technology and having more diverse competencies, both from a disciplinary and a pedagogical point of view [12], p 63. But we cannot wait for today's teachers to have all been replaced by a future generation nor can we hope for such an evolution to happen by itself. Concrete actions thus need to be taken simultaneously both at the in-service level, because some of the teachers actually in place will still be in their classrooms in years from now, and at the pre-service level, to allow future teachers a meaningful contact with the IT environment. Integration of new technologies in schools amounts first of all to integration in teacher education. It is insufficient to inform teachers about "how to use" IT, they must actually work with it. It is only through such an effective use of IT in their own learning that teachers can go through the required change of attitudes that will lead them to perceive the computer as a natural pedagogical tool. But this in turn raises the crucial problem of educating teacher educators.

CONCLUSION

Watson has reported that integration of IT can have "in particular circumstances ... a highly positive impact on children's achievements." [6], p 4. But Anderson and Watson both show that the actual degree of penetration of computers in schools, in particular with respect to mathematics education, is still rather low, with teachers using IT being the exception and not the norm ([2], p 70, [6], p 83). It is thus no surprise that these two recent IEA and ImpacT Studies identify teacher education as a key issue for the coming years. Adequate preparation of human resources being a long-term process, teachers should not delay their personal involvement until a time when some widely accepted "best" solutions to the various pedagogical issues raised by IT might emerge. Fey considers that the actual situation already "offers impressive opportunities for progress." [1], p 266.

Teachers should nevertheless keep in mind that as important as it may be, IT is a resource among other resources, although maybe of a special type. Ruiz reminds us that "... still a large part of their teaching does not involve direct use of computers." [4], p 31. Teaching and learning are first and foremost human activities, involving the teacher and the pupils as well as the interactions between them and with knowledge. IT comes into play to facilitate these interactions.

REFERENCES

1. Fey, J.T. (1989) *Technology and mathematics education: a survey of recent developments and important problems.* Educational Studies in Mathematics **20**, 237-272.

2. Anderson, R.E. (ed.) (1993) *Computers in American Schools, 1992: An Overview.* A National Report from the International IEA Computers in Education Study, University of Minnesota.

3. Kaput, J.J. (1992) *Technology and mathematics education.* In: D.A. Grouws, (ed.), Handbook of Research on Mathematics Teaching and Learning. Macmillan, 515-556.

4. Ruiz i Tarragó, F. (1993) *Integration of Information Technology into Secondary Education - Main Issues and Perspectives.* Guidelines for Good Practice, IFIP TC-3/WG3.1, Geneva.

5. Hodgson, B.R. (1986) *Informatique, mathématiques, éducation et société dans les pays développés.* In: Amara, M., Boudriga, N. & Harzallah, K. (eds.), L'informatique et l'enseignement des mathématiques dans les pays en voie de développement. Actes du 1er Symposium ICOMIDC, Monastir, ICOMIDC and UNESCO, 33-63.

6. Watson, D.M. (ed.) (1993) *The ImpacT Report: An Evaluation of the Impact of Information Technology on Children's Achievements in Primary and Secondary Schools.* Centre for Educational Studies, King's College, London.

7. Howson A.G. & Wilson, B. (1986) *School Mathematics in the 1990s.* ICMI Study Series, Cambridge University Press.

8. Cornu B. & Ralston, A. (eds.) (1986) *The Influence of Computers and Informatics on Mathematics and its Teaching.* (2nd edition) Science and Technology Education Document Series **44**, UNESCO, 1992. 1st edition: Churchhouse R.F. et al., (eds.), ICMI Study Series, Cambridge University Press.

9. K.-D. Graf, (ed.) (1990) *Changing roles of the mathematics teacher in theory and in practice: What are the theoretical changes in the teacher's role?* In: Dubinsky, E. & Fraser, R. (eds.), *Computers and the Teaching of Mathematics: A World View.* Selected Papers from ICME-6, Budapest, 1988. The Shell Centre, University of Nottingham, 28-32.

10. Blakeley, B.H. (1987) *Aspects of computing in a one year secondary mathematics teacher training course.* In: D.C. Johnson and F. Lovis, (eds.), Informatics and the Teaching of Mathematics. Proceedings of the IFIP TC-3/WG3.1 Conference, Sofia. North-Holland, 81-87.

11. Biehler, R., Scholz, R.W., Straer, R. & Winkelmann, R. (eds.) (1994) *Didactics of Mathematics as a Scientific Discipline.* Mathematics Education Library, **13**, Kluwer.

12. Cornu, B. (1992) *L'évolution des mathématiques et de leur enseignement.* In: Cornu, B. (ed.), L'ordinateur pour enseigner les mathématiques. Presses Universitaires de France, 13-69.

13. Dörfler, W. & McLone, R.R. (1986) *Mathematics as a school subject.* In: Christiansen, B., Howson, A.G. & Otte, M. (eds.), Perspectives on Mathematics Education. Mathematics Education Library, **2**, Reidel (Kluwer), 49-97.

14. Yokochi, K. et al. (1993) *Educational features and the role of computers in Japan.* Zentralblatt für Didaktik der Mathematik, **25**, 67-75.

Bernard R. Hodgson received his PhD from the Université de Montréal and since 1975 has been a member of the Départment de mathématiques et de stastique at the Université Laval, where he is now Professeur Titulaire. Beside his research interest in mathematical logic and theoretical computer science, his main professional activities concern teacher education both at the primary and secondary school levels, and in particular the influence of computers and informatics on the teaching of mathematics. He chaired the Canadian National Committee responsible for the organization of the 7th International Congress on Mathematical Education which in 1992 gathered in Québec 3407 participants from 94 countries.

4

An action research role for teachers

Frank (Nick) Nicassio

Information Resource Center
Umatilla-Morrow ESD, Oregon, USA

ABSTRACT

Observations made in four schools during the course of a videodisc instruction project and selected results drawn from a recently conducted survey are used to discuss teachers' roles, responsibilities, and concerns with respect to the integration of information technologies into instruction. These observations and results provide the basis for a proposal to enlarge on teachers' responsibilities to include action research, a role which holds promise for simultaneously promoting effective instruction, improved student learning, and the creation and maintenance of an environment receptive to the constructive introduction of helpful technological innovations.

Keywords: attitudes, collaborative learning, information technology, integration, research

INTRODUCTION

For several years now, the Information Resource Center (IRC) has been integrating information technologies (IT) into various educational activities. The invitation to write this paper has provided an opportunity to reflect on our work, and what we have learned about teacher roles and concerns and their relationship to the delivery of high quality instruction.

Following brief background comments about the context of the IRC's work, I will discuss teachers' roles as observed during an IRC led project to integrate videodisc instruction into four demonstration classrooms. Selected results from a recently conducted survey of our educational service area will be cited to enlarge on the observations. These results and observations motivate a proposal to extend teachers' responsibilities to include action research, a role which holds promise for simultaneously promoting effective instruction, improved student learning, and IT integration.

THE CONTEXT OF OUR WORK

The IRC is a program of the Umatilla-Morrow ESD (UMESD) Student Services Department. UMESD is a regional education centre providing instructional and indirect services to nearly 13,000 students enrolled in 13 school districts in rural northeast Oregon. The UMESD's service area of 66,800 residents consists mainly of small farming and logging communities scattered across 13,500 square kilometres of river valleys, high plateau, and mountains. Our largest school district, located in the Columbia River Basin, enrolls more than 3,500 students in kindergarten through grade 12; our smallest, located in the Blue Mountains, enrolls 50.

Though students are few in number, their needs are as diverse as those of students in many urban areas. During the 1992-1993 school year, for instance, UMESD's Student Services Department instructed over 2,400 Hispanic, Native American, migrant, and at-risk students attending classes in 35 separate buildings. In addition, instruction, therapy, and consultation were provided in the classrooms and homes of nearly 2,500 disabled pre-school and school-aged students.

The organizational and educational challenges presented by geographic remoteness and student diversity have drawn us increasingly to the use of IT to improve student and staff performance. The IRC collaborates with UMESD and local school personnel to integrate IT into selected activities intended to promote student learning and teacher effectiveness. For example, IRC staff have developed computer-based data systems to enhance decision making by multidisciplinary educational teams [1], and to improve multi-agency services for at-risk infants and school children [2]; created a computer-based procedure to improve the quality of psycho-educational reports [3, 4]; established videodisc instruction in middle school classrooms [5]; and have recently begun planning the use of videodisc, CD-ROM, data scanning, and telecommunication technologies to improve the instructional effectiveness of teacher and para-educator teams. Each of these innovations remains operational, the oldest being active for the past ten years.

THE ROLE OF TEACHERS

Teachers introducing a technological innovation
The IRC's most recent operational endeavour in IT integration began in 1990 when UMESD funded videodisc demonstration sites within four of our 13 school districts with the goal of helping teachers provide high quality mathematics and science instruction to mildly handicapped, non-handicapped, and talented and gifted students, who are educated in the heterogeneously grouped classrooms characteristic of our rural area. Videodisc (laserdisc) technology promised to be one means by which teachers within established classrooms could adapt existing curriculum and their own instructional methods to meet a more diverse array of learning needs for students of 11 to 15 years of age.

Our initial choice of instructional laserdiscs was the "direct instruction" oriented Core Concepts in Mathematics and Science series which was validated for the National Diffusion Network by its principal developer, Utah State University. We subsequently added The Adventures of Jasper Woodbury, an "anchored instruction" oriented series which was, at the time, under development by the Learning Development Center at Vanderbilt University. Teachers have since integrated both programs into classrooms where they now serve approximately 705 students annually with generally favourable learning outcomes [5].

I believe that the high quality of laserdisc instruction contributed in large measure to our success. I also attribute success to the roles teachers assumed as they wrestled with the complexities associated with introducing a technological innovation into established classroom routine. We had approached the integration of instructional videodisc in a planned and systematic manner hoping to deepen its penetration into daily classroom practices, amplify its impact on teacher enthusiasm, and increase its effect on student performance. But, our efforts were also motivated by a desire to have teachers' roles emphasize collaborative inquiry and action. Why? Because it is these qualities, I believe, that aid in creating and maintaining an educational environment receptive to the constructive introduction of significant changes, including instructionally effective technological innovations. Let us review the process by which we attempted to foster these qualities.

A process to foster collaborative inquiry and action

We established the four demonstration sites on the recommendation of a two-county co-ordinating council. Following approval by local district superintendents, we solicited potential sites with a "request for proposals" distributed to each school in our service area. District co-ordinating council representatives responded to initial queries from schools. Four sites submitted proposals to the IRC which subsequently made visits to show a videotape featuring laserdisc instruction, to answer questions and clarify goals, and to emphasize that criteria for selection required a school to establish a team responsible for both videodisc instruction and evaluation.

The IRC then asked each of the four school sites to decide on team membership and to discuss team options for effectively integrating videodisc technology into instructional programs, a step which took two months to accomplish. The four resulting teams varied in size and role composition, but consisted of at least two regular classroom teachers, one special education teacher, and the building principal. Other personnel that were also involved included: a special education director, a speech and language therapist, a student services coordinator, a parent, a curriculum consultant, and two media specialists.

Our initial assumptions about the importance of establishing "teams" was confirmed in June 1993 by an IRC survey of educators (teachers, principals, para-educators) who have been actively engaged in IT integration efforts throughout our service area. In that survey, every respondent (N=21) acknowledged that, in addition to teachers, other personnel were crucial to the integration and effective utilization of information technologies. The survey emphasized the importance of a wide range of personnel, including media specialists, resource specialists, classroom assistants, technicians, technology specialists, principals, assistant superintendents, secretaries, school boards, universities, researchers, parents, students, teacher teams, and school-governance councils. These persons were judged important because they provided access to resources, enhanced the integration process, applied technology directly to student learning, and added a new air of collaboration to the educational process.

Although teachers have special instructional responsibilities obliging them to focus IT efforts on teaching and associated activities, our survey data suggest that they are functionally dependent upon support personnel to realize their objectives. Consequently, it becomes imperative for teachers and related education personnel to collaborate in their efforts to organize IT in valuable and productive ways.

Following the preliminary discussions cited above, each school team prepared a draft of a First Year Implementation Plan which was submitted to the IRC for review. The IRC then provided hardware, software, training, and a two-day field trip to a neighbouring state so that teams could directly observe laserdisc instruction in classrooms.

Upon return, teams wrote their final First Year Implementation Plan which, among other things, specified: the names of the school's team leader, each team member, and a research journal keeper; videodiscs, adopted texts, and enrichment materials to be used for instruction; and the standardized and informal assessments to be used for evaluating laserdisc effects. They also included how the team would structure staff responsibilities, class assignments, and instructional time to insure active student participation; a description of three adaptive curriculum strategies to be used in conjunction with videodiscs for the benefit of disabled students; and supportive assistance the team would require from the IRC.

Formulation of the Implementation Plan not only committed each team to a collaborative and systematic trial using laserdisc instruction, but also to a collaborative and systematic evaluation of the effects of that instruction on teacher perceptions and student learning. Under the plan, each site would submit its findings to the IRC in the form of follow-up questionnaires, formal and informal student performance measures, a research journal, a written narrative summarizing all data and anecdotal observations, and a modified Implementation Plan for the subsequent year.

Over the four years of the project, the IRC has invited sites to an annual, day long, working meeting held at the beginning of the new school year. These meetings have provided the means for teams to share their findings, discuss their experiences and insights, suggest modifications for successive instructional trials, and recommend additional training. Results have also been shared at local, regional, and international meetings. Team recommendations for additional software and training have led to the acquisition of other videodisc programs, the offering of college credit for practicum experiences, and a workshop on multimedia technology.

The above portrayal should not be interpreted as a smooth and flawless integration, however. It was not. Teachers always operate in a complex environment of interdependence between instruction, technology, and human relations. Our implementation was no exception. As a consequence, we experienced periods of higher and lower motivation, diminished commitment, discontinuities resulting from staff turnover, technical problems, and difficulties in correlating instructional objectives presented on videodisc with objectives presented in textbooks and in state curriculum guidelines. Also, extending the program beyond model sites has progressed more gradually than anticipated.

What we created, however, was a socio-technical environment in which teacher roles and responsibilities for instruction were sustained by a larger constellation of administrative support, technical assistance, in-service, cooperation within and among teams, and IT utilization. I believe it is exactly this kind of collaborative environment that can assist teachers as they labour over the many complex problems they confront daily. Our June 1993 survey revealed, for instance, that teachers' IT integration concerns clustered around three themes:

- barriers to full access and full utilization of existing technologies;
- lack of personal preparedness;
- technology's potential for unanticipated negative consequences.

Each of these areas of concern could be constructively addressed in an environment that provides administrative support, technical assistance, in-service, and active cooperation.

In addition, we also established an action research context in which teams defined input (independent) and output (dependent) variables, specified outcome measures, modified successive trials based on empirical evidence, and shared their findings with other professionals. Callister and Dunne have noted that technology dazzle can induce teachers into accepting instructionally ineffective software packages" ... even if it means denying the evidence of personal experience" p. 325 [6]. To overcome this tendency, educational practices must encourage teachers to make data-based decisions and mid-course corrections in order to modify potentially effective technologies for the benefit of effective student learning. I believe the adoption of action research represents an important step toward fostering to a greater degree than currently practised, these conditions of collaborative inquiry and action.

A PROPOSAL TO EXTEND TEACHER ROLES TO INCLUDE
ACTION RESEARCH

Kemmis and McTaggart [7] describe action research as a process by which members of a school community (teachers, administrators, students, parents, school boards, support personnel) can work collaboratively toward improvement and change in their schools. There are variations on the action research theme, of course [7 - 13], but most provide for thinking systematically about what happens in schools, implementing critically informed actions, and monitoring and evaluating the effects of the action with a view towards continuous improvement and ongoing collaboration with others who share a common concern [7]. Action research may, thus, be described as a process of collaborative thinking, action, and adaptation for the purpose of improving teaching and learning. It may also be thought of as a process by which schools can collaboratively adapt information technologies to attain their important educational goals and instructional objectives.

The American passion over the last decade for educational reform and restructuring [14] has, coincidentally, set the stage for teachers and related personnel to assume the collaborative action research role proposed by this paper. Federal legislation such as the Technology for Education Act of 1993 [15] with an interest in technology diffusion in schools, and state legislation such as Oregon's Educational Act for the 21st Century [16] with its mandate for decentralized school-based decision-making, should make it increasingly possible for a growing number of teachers, administrators, and para-educators like those we surveyed to assume an active role in determining effective uses of information technologies in the instructional, administrative, and organizational activities of their schools.

How might educators join technology and education together in a productive discussion? By adapting Sirotnik's dialectical inquiry-action cycle [13], school-based action research teams could ask the following questions and take the following actions.

• What are our current beliefs and practices with respect to student learning experiences (or professional technical training, or instruction in literature, or governance of the school site, or gender equity in mathematics and science instruction, or full inclusion of disabled students in mainstream classrooms ...)?
• How did we come to hold these beliefs and employ these practices?
• Whose interests are, and are not, being served by these beliefs and practices and in what ways?
• What information and knowledge do we have or need to get hold of that may illuminate this discussion about beliefs and practices?
• Is the current state of affairs the way we want it to be?
• What are we going to do about it?
• Can information technologies extend our capacities to achieve our proposed actions? If so, how can we incorporate their use into our plan?
• Take actions relevant to our plan, monitor these actions, and then repeat this inquiry-action cycle.

Participants are encouraged to invite a small group of colleagues to select a technology issue of interest and ask the questions contained in this inquiry-action cycle in order to better grasp the complex interdependencies between instruction, technology, and human relations confronting teachers on a daily basis. This little experiment may help a consideration of how teachers and related support personnel might clarify the circumstances under which they must work and learn together in order to create and maintain an environment receptive to the constructive introduction of helpful technological innovations. It may also suggest the salutary effect that collaborative inquiry can have on the successive trials of an implementation.

Technologists may complain that collaborative inquiry and action are irrelevant to the integration and effective use of information technologies. I would contend, however, that ignoring the sort of dialogue stimulated by Sirotnik's inquiry-action cycle is precisely the reason that, "... over the past dozen years ... technology has not revolutionized learning in the classroom, nor led to higher productivity in schools" [14], p.188. By not subordinating our interests in technology to a larger discussion about the socially legitimate work of education, and how that work is to be sustained in the classroom on a daily basis, we have diminished IT's potentially beneficial impact on schooling. Rhodes of the American Association of School Administrators has noted, "Sadly because of these flaws in our perception of schools, the 'work' they do and how they do it, we have today numerous examples of technology's effectiveness with real children, in real classrooms. But without a context within which to make sense of them, they appear as isolated demonstrations of technology" [17], p.3. Action research can provide teachers "a context within which to make sense"; a context of joint inquiry, mutual consensus building, collective planning, concerted action, and the generation of empirical feedback against which to evaluate current efforts.

Enthusiastic collaboration among all participants (teachers, principals, para-educators, school boards, parents, students) in the educational process is the key to defining and solving problems about daily educational practices. Only in this constructive atmosphere can specific technologies be modified, over successive trials, to address desirable goals. Adoption of collaborative problem framing and collaborative action taking, of the type defined by Sirotnik's inquiry-action cycle, would mark a new and emerging role for teachers, and would clear the way for the genuine integration of information technologies into daily educational practices. It would also alleviate many of the teacher concerns identified by our survey.

In conclusion, we must acknowledge that some people will take the position that technology is an essential yet only small part of the education equation [18], while others will take the more radical stance that economic progress is dependent on the rapid and total replacement of conventional schooling by a new commercial industry based on technology [19]. Whatever stance political pressures push education in the direction of, an action research role has potential for enabling teachers and the related personnel upon whom they depend to simultaneously promote effective instruction, improve student learning, and integrate information technologies into daily educational practice.

Acknowledgments
I would like to thank Dr. Robert Earl, Dr. Dean Thompson, William Miller, Suann Ritchie, and William Taylor for their critical comments of an earlier draft of this paper. Thanks also to Karen Overstreet, Tricia Towne, and Karen Hoefl, the staff of the Information Resource Center who make its work possible.

REFERENCES

1. Nicassio, F.J., & Overstreet, K.K. (1993) *A total picture of the child.* Educational Information Resource Management Quarterly, **2** (3), 18-22.

2. French, R.B., Overstreet, K.K., & Nicassio, F.J. (1991) *Community networking, early identification, and electronic information systems.* In, The Oregon Conference Monograph, 159-165. Eugene, OR: University of Oregon.

3. Nicassio, F.J., & The Staff of The Child Development Program. (1986, Spring) *Allegro: A system for creating psychoeducational reports using a word processor knowledge base.* Computers in Psychiatry/ Psychology, **8** (1), 5-11.

4. Nicassio, F.J., Moore, K.J., & The Staff of The Child Development Program (1986, July) *Computer support in a rural setting: A comparison of computer-assisted and handwritten evaluation reports.* Psychology in the Schools, **23**, 303-307.

5. Nicassio, F.J., Thompson, D., Burk, J., Wright, K., Palmer, N., McBride, M., & Shubert E. (1993) *Laserdisc Technology in the Rural*

Classroom: A Case Study of Anchored Instruction in Mathematics and Science. A. Knierzinger, & M. Moser (Ed.), Informatics and Changes in Learning: Proceedings of the IFIP Open Conference. Gmunden, Austria. International Federation for Information Processing, XVIII:1-4.

6.   Callister, T.A., & Dunne, F. (1992) *The computer as doorstop: Technology as disempowerment.* Phi Delta Kappan, 324-326.

7.   Kemmis, S., & McTaggart, R. (1988) *The action research planner.* Geelong, Victoria: Deakin University Press.

8.   Goswami, D., & Stillman, P. (1987) *Reclaiming the classroom: Teacher research as an agency for change.* Upper Montclair, NJ: Boynton/Cook Publishers, Inc.

9.   Joyce, B., Wolf, J., & Calhoun, E. (1993) *The self-renewing school.* Arlington, VA: Association for Supervision and Curriculum Development.

10.   Mohr, M.M., & Maclean, M.S. (1987) *Working together: A guide for teacher-researchers.* Urbana, IL: National Council of Teachers of English.

11.   Nicassio, F.J. (1992, January/February) *SwampLog: A reflection in action journal process.* The Writing Notebook, **9** (3), 13-18.

12.   Reason, P. (Ed.). (1988) *Human inquiry in action: Developments in new paradigm research.* Newbury Park, CA: Sage Publications.

13.   Sirotnik, K.A. (1987) *Evaluation in the ecology of schooling: The process of school renewal.* In J. Goodlad (Ed.), The ecology of school renewal. Eighty-sixth Yearbook of the National Society for the Study of Education: Part 1. Chicago, IL: University of Chicago Press, 41-62.

14.   Conley, D.T. (1993) *Roadmap to restructuring: Policies, practices and emerging visions of schooling.* Eugene, OR: ERIC Clearinghouse of Educational Management, University of Oregon.

15.   SB1040 (1993) *Technology for Education Act of 1993: United States Senate Bill 1040: Executive Summary.* Educational Information Resource Management Quarterly, **2** (4), 30-34.

16.   HB3565 (1991) *Oregon Educational Act for the 21st Century: House Bill 3565.* The 66th Oregon Legislative Assembly, Regular Session.

17.   Rhodes, L. (1993) *What to do until technology legislation arrives.* Arlington, VA: American Association of School Administrators.

18. (1992) *School reform: Why we need technology to get there.* Electronic Learning, 22-28.

19. Perelman, L.J. (1992) *Hyperlearning: Clinton's greatest opportunity for change.* Discovery Institute Inquiry. Seattle, WA: Discovery Institute.

Nick Nicassio is a resource specialist for the Umatilla-Morrow Education Service District where he coordinates the Information Resource Center. Nick received his PhD in Educational Psychology from the University of California, Santa Barbara, and serves as adjunct faculty member for Oregon State University Division of Continuing Higher Education. He is a member of the American Educational Research Association, the American Psychological Association, the American Psychological Society, the Association for Humanistic Psychology, the National Society for the Study of Education, the International Society for Technology in Education, the Association for Supervision and Curriculum Development, and Phi Delta Kappa - Eastern Oregon Chapter.

5

Classroom ethos and the concerns of the teacher

John Olson

Queen's University
Ontario, Canada

ABSTRACT

It is useful to think of information technology as representing a cluster of educational ideas which may or may not challenge existing practices. How teachers respond to those challenges constitutes the potential for improvement.

Superficial accommodation to technology is everywhere evident and reported, but what are the deeper implications for teachers of particular challenges posed by IT to how subjects are construed and taught? This paper will consider research which illuminates increasingly deeper concerns of teachers about IT and the classroom ethos.

Keywords: classroom practice, curriculum development, innovation, integration, pedagogy

INTRODUCTION

In this paper three major aspects of integration will be considered. First, I will consider IT and the process of change; for example, integrating IT into classroom routines, rewriting curriculum materials, researching software, restructuring the classroom layout. This leads secondly, to integrating with and reflecting on the pedagogical theory underlying the IT "package" or cluster of ideas; the role of teacher in using IT, the nature of student activity, the nature of subject matter knowledge, the content of the activity, the connection of classroom work to the world outside. Finally, I will consider the implications of teacher concerns for research and development of IT.

IT AND THE PROCESS OF CHANGE

Watson [1] notes that where computers are located is an important issue for integration. She says that teachers having to rely on specialist computer rooms may be inhibited in integrating computers. She locates the problem in the difficulty of access and in the different ways of working which computer managers and teachers might have. These factors can create a distance between teacher and class. She says there is also loss of control which can undermine confidence. She concludes that micros should be located in subject-based rooms which provide a supportive environment for teachers, and that successful integration depends on such an environment. She refers to other work which supports this idea: "unless we consider more carefully the nature ... (of) classrooms themselves we cannot expect that our software development and in-service activities will have any more impact in the future than they have in the past" p. 36. The computer is one tool amongst others found in classrooms, not a classroom in itself. Watson is urging that the computer be seen as part of a complex ecology of resources which lie at hand for teachers in tacit ways.

Tanner [2], reporting on a study of the implementation of IT in mathematics education, found that access to computers is a major problem for teachers. The access problem could be alleviated if computer studies courses were done away with, Tanner recommends. He suggests that "activities which develop IT skills through subject teaching should be built into the scheme of work of all subject areas" p. 148. He suggests that computers studies teachers become advisors to those using computers in subject teaching.

Van den Akker et al [3] note the extensive literature on implementation which points to the need for a developmental approach to change. They call such an approach to integration of computers "an infusion approach". It is a staged approach in which early success is vital if implementation "fidelity" is to be achieved. Early success, they believe, can be achieved if the materials reflect an awareness of the problems teachers face in implementation of innovations. Problems they point to are: lack of time, background knowledge and appropriate "didactical role" p.73, and restricted sense of the learning potential of the material. An initial study, in which materials were designed with these problems in mind, suggested that high fidelity implementation could be achieved when teachers had initial success with materials.

Voogt [4], a colleague of van Akker, notes that, even though IEA data for the Netherlands suggest increased computer use in science education, data about the intensity of use are not available, and more needs to be known about the implementation process. She points out that even though concrete guidelines for use of software are provided teachers do not necessarily act in accordance with those guidelines.

In her research she found that teachers did not use the teacher guide designed to accompany a science programme. Even though the 'operational manual' was placed in their hands, teachers, used the software according to their own planning. As she notes, "Secondary science teachers already have their own subject matter and pedagogical knowledge. In their perception, therefore, they probably felt less of a need for special support" p. 144. However, she suggests that teachers may not properly exploit the curriculum potential of the software through failing to "understand it" .

Papert [5] also addresses the role of teachers in change. Like Van Akker et al he sees changes as a process of overcoming obstacles. The obstacle that Papert had addressed in the past was the teacher. He now sees the school organization as the obstacle, and teacher the victim. Unlike Akker et al, he does not think that changes in school management can lead to realizing the potential of computers: "School's hierarchical organization is intimately tied to its view of education, and in particular to its commitment to hierarchical ways of thinking about knowledge itself" - p. 61.

Eraut [6] summarizes research on implementation problems. He says there are close parallels between IT in education and earlier innovations on the issue of clarity of purpose. He notes that teachers often have real doubts about the extent to which pupils will benefit from innovations. Even if teachers do incorporate computers in their teaching, Eraut notes, they reach only a certain level of use. To go beyond that level, help in reflecting on the process is needed if computers are to transform the curriculum. The assumption is that computers have transformative potential.

Miller and Olson [7] take issue with this idea. Based on their review of the implementation literature and their own research, they conclude that, on the whole, computers amplify rather than transform practice. They note "Expert systems, virtuality and hypermedia are new technological areas and the object of interest of futurists, however, those educators who make the best use of these advancements may be those who look backward -- or at least look around -- before leaping ahead" p. 138. In their research they document instances where pre-existing innovative practices are amplified by computers, but where traditional practices are not transformed by using computers.

The process of change thus involves the interaction between new technologies and pre-existing practices, as we have seen. The second aspect of integration concerns reflective practice.

IT AND REFLECTIVE PRACTICE

Ragsdale [8] reviews research on classroom use of computers. He notes that "In classrooms, it seemed particularly difficult to make the use of computers part of the classroom while preserving the teacher's style" p. 682. Teachers appear to be less critical about and demanding of student performance when at the computer. He notes that teachers appear uncritical in their assessment of how their students are using computers. Students are given approval for using computers in trivial ways; approval that they would not receive were they using other learning tools. He draws a distinction between using computer tools and applying them. Application, in his view, means using computers to discover relationships - to use them to see things in new ways. Doing this depends on a culture of inquiry already existing in the classroom. Such a culture, he says, "ultimately depends on relationships between people" p. 683. He re-iterates the concern about the classroom as a context in which computers are used and the importance of that context for achieving IT potentials.

Upitis [9] points out the danger of contrived learning, in which uses of technology are "contrived by teachers in order to somehow make use of available technology" p. 233. Lack of easy access that Watson referred to, may itself give rise to contrived teaching, as a teacher coping strategy. Upitis says that, for there to be reality in electronic communication there has to be a feeling that students are dealing with real issues: "Electronic mail systems are best used when there is a need to learn the tool." Implicit in this idea that the tool is at hand. So the issue of contrived learning, that is to say the quality of work, is a function of both needing the tool and also having it, and having it means having it "to hand".

Brown [10] reviews research on the implementation process in relation to an evaluation of the UK Information Technology Teacher Training Development Programme (ITTTDP). He argues that teacher concerns about IT are evolving rapidly and that these concerns are the subject of reflection and clarification. The need for opportunities for teacher reflection was stressed. The teacher training material developed by the ITTTDP asked teachers to reflect on the learning outcomes supposedly enhanced by the use of IT. In this way the concern about trivial adoption raised by Ragsdale was addressed. Brown also points to the importance of the overall context for learning when assessing the potential of IT. Brown notes that IT can be a stimulation for change "opening up the possibility of transforming aspects of the teaching and learning environment" p. 149. He points to the speed at which results can be generated, freeing up time for discussion and thus deeper understanding. How the tool influences the teaching is, of course, an ongoing issue. Teachers will need to be able to lead effective discussions if this promise is to materialize.

The Council of Europe report Education and the Information Society, edited by Eraut [6], has a chapter on approaches to implementation in different member countries. The report notes that "Both the teacher and the technology have to adapt to each other" p. 63. Olson [11] makes the same point: "The ideas built into software ... can give rise to more intelligent form of teacher-student relationship, but only if software designs are based on an understanding of school life" p. 123. Watson [12], for example, sees teachers as key members of software development teams.

ISSUES FOR DEVELOPMENT OF IT IN EDUCATION

Such research raises questions about the risks teachers run in extending their ways of teaching to the edge of their comfort zone. These risks involve fundamental issues for teachers, because as Olson [11] states, they are interlinked with the way they are perceived by their students, parents, and administrators and with their own self-concept as teachers.

At the heart of teacher concerns is the maintenance of a certain ethos in the classroom. Those who design software need to do so keeping in mind the concerns of teachers. To better understand integration of IT it is useful to think in terms of two dimensions of change - the instrumental and expressive dimension, and the accommodation-assimilation dimension.

Instrumental actions are meant to achieve certain outcomes; they are intended to be causal. Expressive actions are meant to make statements about what things mean; such actions express values. Teacher use of IT is both instrumental and expressive, and the expressive use of IT often precedes the instrumental and may be at odds with it. The use of this distinction is necessary in order to understand why teachers are willing to make the often heavy adjustments needed to achieve even minimal integration of IT in the classroom, and to appreciate the satisfaction and frustration they experience in this process.

Accommodation and assimilation refer to the demands made on teachers by integration. There are the practical problems of accommodating computers into classroom routines. Considerable research on integration addresses such problems. But then there is the process of reflecting on these changes; the process of assimilating the new arrangements within the schemes teachers have for teaching and learning. Accommodating IT challenges those schema. These two dimensions give rise to the ethos of the classroom.

Why is it important for IT designers and managers to consider the ethos of the classroom: that is to appreciate the purpose and methods of those who are attempting to integrate IT in their classrooms? Why is it important for software designers to make explicit their ideas and show how they belong in schools? Teachers can learn much from the ideas of software innovators. Linking teachers and innovators together in innovation takes change as process of dialogue rather than one of implementation. Without this process software development will not lead to a better ethos in schools.

CONCLUSION

The research I have reviewed suggests that computers in the classroom have many potentials. They perform complex transformations of information rapidly and offer multiple representations of information. They let students probe the systems in order to understand their structure and help them to get to the bottom of things. They form bridges between learning different but related things.

Computers may also be important because they help teachers confront their experience - to make sense of it and thus become aware of the value of what they do. Change is a process of reassigning value to practice as well as actually adopting new practices.

Similarly, it is not only the cognitive capacity of students that is at issue when they tackle problems posed to them by computer programs. They are being asked to see the world in new ways, to value things differently. How they do this depends on prior experience. As outsiders we need to know the network of "ideas" which teachers and students bring to their practices. We need to know much more about the nature of existing practices in schools as a context for understanding how IT is integrated. Change, as has been said already, is not just a matter of adopting new technologies - of being innovative - it is coming to teach and learn in new and more productive ways, and this involves the challenge to reflect on practice that IT represents. Such a challenge is the great potential of IT.

REFERENCES

1. Watson, D. (1990) *The classroom versus the computer room.* In: M. Kibby (Ed.), Computer assisted learning. Oxford: Pergamon.

2. Tanner, H. (1992) *Developing the use of IT within mathematics through action research.* In M. Kibby and J. Hartley (Eds.), Computer Assisted Learning, Oxford: Pergamon Press.

3. van den Akker, J., Keursten, P., & Plomp, T. (1992) *The integration of computer use in education.* International Journal of Educational Research, **17** (1), 65-75.

4. Voogt, J. (1993) *Courseware for an inquiry-based science curriculum.* The Hague: CIP Publishers.

5. Papert, S. (1992) *The children's machine.* New York: Basic Books.

6. Eraut, M. (1991) *The information society: a challenge for education policies?* In: M. Eraut (Ed.), Education and the information society. London: Cassell.

7. Miller, L., & Olson, J. (1994) *Putting the computer in its place: A study of teaching with technology.* Journal of Curriculum Studies, **26**, 121-141.

8. Ragsdale, R. (1992) *Making computer tools effective in classrooms.* Proceedings of the Ninth International Conference on Technology Education. Austin, TX: University of Texas.

9. Upitis, R. (1990) *Real and contrived uses of electronic mail in elementary schools.* In M. Kibby (Ed.), Computer assisted learning. Oxford: Pergamon Press.

10. Brown, A. (1994) *Processes to support the use of information technology to enhance learning.* Computers and Education, **22**, 145-154.

11. Olson, J. (1992) *Trojan horse or teacher's pet?* In: Computers and the teacher's influence. International Journal of Educational Research, **17** (1), 77-84.

12. Watson, D. (1987) *Developing CAL.* Computers in the curriculum. London: Harper and Row.

13. Olson, J. (1988) *Schools/Microworlds, Computers and the culture of the classroom.* Oxford: Pergamon Press.

John Olson works with pre-service teachers in science education and does research and graduate work in the field of curriculum change. He is particularly interested in understanding the teacher's point of view in the innovation process. Over the last 10 years he has conducted a number of research projects on the integration of computers in the classroom. His recent book *Understanding Teaching* is mostly based on that research.

6

Using evidence about teacher development to plan systemic revolution

Jim Ridgway
Don Passey

Department of Psychology,
University of Lancaster, United Kingdom

ABSTRACT

In England and Wales Information Technology (IT) is to be used to support the teaching of all subjects; in addition, all students have an entitlement to IT capability. Data from several surveys are reported, which do not match this vision of IT. Rather, IT is seen by teachers to be concerned with the acquisition of technical skills, or as a support for administration: a small number are terrified of IT, rather more are unconvinced of the benefits of IT, and only a minority of teachers use IT to support their teaching. Teacher concerns regarding IT are reported. Finally, the paper offers a model of school development which highlights the changing demands on teachers as IT is integrated progressively into the curriculum.

Keywords: innovation, management, integration, case studies, attitudes

INTRODUCTION

The main challenges we face when attempting to integrate Information Technology (IT) into educational practices are concerned with creating and maintaining an educational climate where steady change in teacher knowledge and classroom practice is accepted as a desirable state, planned for, and supported appropriately. These changes are not simply changes of content - a shift from studying the Romans to the Renaissance, or swapping statistics for algebra - rather, they pose deep questions about the most useful things students can learn in school, and about the uses of the most powerful intellectual supports we have.

In any change process, it is essential to have a view of where different key agents are, in terms of both their actions and beliefs, and to plan changes on the basis of this information. Change is not just a matter of skill development, or the reform of individual attitudes - change needs to be embedded in a nurturing institutional context. It is clear that individuals can develop on their own, but for IT to be integrated across departments, and the school curriculum, a broad range of institutional issues must be addressed, which include the provision for the development of individuals who do not necessarily share the institutional vision of IT.

This paper begins by setting the scene in England and Wales - where a national curriculum has been introduced which requires that students learn using IT, and become fluent users of IT itself. It then presents evidence gathered from case histories of schools, survey data, and interviews. Surveys of teacher attitudes, and structured interviews show that a large proportion of teachers make little use of computers in their teaching, and that a significant number actually express a fear of the new technology. Of the teachers who do make use of IT, many make poor use, and misperceive the rationale for IT in education. Schools differ a great deal in the extent to which IT is integrated into their curricular practices, and surveys and case studies show a wide variety of perceived and actual issues (for example, software choice, equipment deployment, phasing of staff development) which schools need to address.

We offer an account of the stages of IT development in schools and how this relates to teacher development, and describe the approach used by the STAC project to support the integration of IT across the curriculum.

CURRENT IT USE IN EDUCATION

The National Curriculum in England and Wales is specified largely by subject (for example, Science [1], and English [2]). Statutory Orders set out Programmes of Study which must be followed, and define Attainment Targets (ATs) which should be reached, in each subject; these ATs refer to different domains within each subject, such as Shape and Space, and Handling Data, in Mathematics [3]. They list things students can do, in ascending order of difficulty. References to the use of IT in teaching and learning appear in every one of the Statutory Orders; so every student is to receive teaching which is based on IT in every subject in the curriculum, and it follows that all teachers are expected to be able to use IT as a teaching aid. IT capability is also defined as a subject area within the Technology Statutory Orders [4]. This is a broad conception of IT literacy which sets out ATs under the headings of: Handling Information; Communicating Information; Modelling; Measurement and Control; and Applications and Effects. The National Curriculum, then, identifies two distinct roles for IT; one is as an end in itself, to empower students by developing their abilities to use IT as a tool whenever it is needed; the other is as a support for learning in every curriculum subject.

What changes in IT provision are apparent?

What progress has been made so far? Schools have had just 3 years since the introduction of the National Curriculum. Some idea about the changes in the sorts of activities which can be supported via IT come from a recent evaluation into the effectiveness of the computer support system in one Local Education Authority [5]. In 1988, the 44 secondary schools had 782 computers, and in 1993, they had 2207. The computer - student ratio fell from 1:45 in 1988 to 1:14 in 1993 (the best equipped school had a ratio of 1:5, the worst was 1:25). There has also been a change in patterns of purchase, with a steady increase in both the number of laptop computers and printers being purchased, a steady increase in the use of the LEA library of CD-ROMs, and very large increases in the number of blank disks being purchased (from 50 in 1989-90 to 3080 in 1992-93).

Of course, computers and software packages have increased dramatically in power over recent years, and continue to do so. Technical innovations such as laptop computers and CD-ROM offer educational opportunities which we are just beginning to explore, but which offer the potential for new ways of working in education.

In order to make good use of these developments, a climate of experimentation, and acceptance of short term failure, the use of methods to promote the rapid dissemination of effective educational practices, and a willingness to accept steady change as a way of life, will be essential. Such climates are difficult to create. In order to begin the process of change, it is necessary to have some knowledge of teacher beliefs, their knowledge of and current classroom uses concerning IT.

What do teachers believe IT is good for?

An unpublished survey of 54 primary school teachers by Gower [6] explored teacher knowledge of the National Curriculum, and beliefs about the benefits which can accrue from IT. Twenty six out of 54 teachers were unable to name correctly a single skill contained in the Statutory Orders for IT. Eleven out of 54 teachers identified skills which are not there at all: notably skills relating to keyboard skills, co-ordination and movement. So the idea of IT capability as a set of general purpose skills which can enhance personal effectiveness is not well understood: rather, IT is seen as a set of technical skills.

Teachers were asked about the benefits which accrue from the use of IT. Open format responses were then categorised under different headings. The number of teachers spontaneously offering remarks under each heading are shown in Table 1 below.

Benefits of IT use	% Respondents
Low level cognitive skills	52
High level cognitive skills	33
Personal skills	30
Social skills	17

Table 1: Percentage respondents who indicate particular benefits of IT use (n=54) (Gower, 1992)

These data support the earlier conclusions; the dominant use of IT is to develop technical, rather than conceptual, skills.

The STAC Project offers a consultancy service to schools. Consultancy consists of an analysis of current human and physical resources (elicited via a questionnaire to staff, structured interviews with about half of the school staff, a review of the school prospectus and other materials they have produced, and direct observation of classroom practices), and recommendations for future action. Consultancy is an essential part of our work - while the primary goal is to offer direct advice to individual schools, a secondary goal is to keep STAC in touch with current teacher concerns about IT and the state of school development in general. To share some of this knowledge with a broader community, we publish some consultancy reports, with the permission of the school concerned. Of consultancies conducted so far, 3 are available in published form [7], [8], [9]. Data from these reports are presented below.

In one school, teachers were asked about the uses they saw for IT. Their responses were categorised and are shown in Table 2 opposite.

Teacher Perception of IT	Number of teachers
Administrative tool	18
Teaching tool	16
For skill development	14
Support learning	12
Material production & presentation	10
Games, or for playing	9
Something to know about	5
Motivates children	5
Bandwagon	3
Expert tool	2
For programming	2

Table 2: Teacher perceptions of IT from an interview survey with 66 staff in one school (n=66) (Passey and Ridgway, 1992)

The commonest single response was to see IT as an administrative tool. Overall, the responses were rather sparse - for example, there were only 46 responses from 66 members of staff which related to educational uses. Further, 16 responses are concerned with what the teacher does rather than what the student does, 14 responses refer to skill development, and 4 responses are unrelated to the National Curriculum view of needs for developing IT Capability. Only 12 responses are concerned with the support that IT can offer for learning. These data show that in this school, educational ambitions for IT are somewhat modest.

In two schools, staff were asked to categorise their use of IT on a scale which ranged from 'IT terrified' to 'IT curriculum user'. Their responses are shown in Table 3 below.

How teachers viewed their use of IT	Report 1* n=66	Report 2** n=53
IT terrified	5% (3)	10% (5)
IT unconvinced	20% (14)	8% (4)
IT convinced (but not using IT in lessons)	30% (20)	28% (15)
IT administrative user	25% (16)	51% (27)
IT curriculum user	45% (30)	51% (27)

Table 3: How Teachers in Two Schools Viewed Their Uses of IT (* from [7], data collected 1991; ** from [8], data collected 1991)

Teachers who are terrified of IT need careful nurturing; those who are unconvinced need some persuasive communication, those who are convinced, but who do not use IT in lessons, need vivid exemplification, resources and encouragement, and curriculum users probably need encouragement to push forward their IT uses, as well as encouragement to help others.

Experience reported	No. of respondents
Inexperienced	19
Some experience	16
Experienced	4

Table 4: The Way in Which Teachers Reported their Level of IT Experience in One School (n=39) (Passey, 1994)

Classroom Use	No. respondents
Used at least one per week	5
Used occasionally	21
Used never	11

Table 5: The Regularity of Classroom Use of IT Reported by Teachers in One School (n=39) (Passey,1994)

Similar evidence was gathered from a third school consultancy, and data are shown in Tables 4 and 5. The modest use of IT in classrooms can be compared with statistical evidence from the Department for Education (DfE) [10] which reported in 1992 that 32% of all teachers in all secondary subject departments were using a computer in class at least twice each week. DfE data contrasts with data from the ImpacT study [11] where just 12% of teachers used a computer at least once per week. School consultancies produce a long list of teacher concerns about IT. Some of the major concerns are listed here:

• beliefs about why IT is being introduced into education (e.g. a misplaced belief about the needs of employment, a plot to save teacher salaries) and what IT is really good for, in educational terms (e.g. playing games, and drill and practice);
• concerns about change itself, such as the number of changes taking place at the same time, and the uncertain nature of government planning - will these changes still be required in a few months?
• time pressures;

- loss of traditional skills, such as handwriting, spelling, and mental calculation;
- IT can be seen to be irrelevant to the teaching of their subject, which can be done better, and more cheaply, in other ways;
- quality of software, and its match to the existing curriculum;
- personal confidence and competence;
- inversion of roles, where students know more than teachers;
- resource implications - direct and indirect costs, and the need to replace equipment regularly;
- problems of access by staff and students;
- security; the problems of vandalism; and file security for the work of students and teachers;
- electrical safety;
- worries about radiation associated with prolonged working with screens; and
- space, and a loss of flexibility in room uses.

What sort of support is appropriate?
Teachers are always asked how they learn most effectively. In School 1, 45 reported that it was by doing, 19 indicated that it was by working alone, 16 by discussion and talking, 15 by working in a group, 10 by reading, 10 by observing, and 8 by listening. If self report matches actual learning style, these data suggest that most teachers would benefit from using IT in their classrooms or having direct access to it, and from group work and discussions. An innovatory feature of teacher support in England and Wales has been the provision of advisory teachers, who work alongside other teachers in class. This sort of support is expensive to provide, but seems well suited to teachers' self-reported needs.

When questioned about perceptions of themselves as teachers, 39 teachers felt they acted as a facilitator; 30 indicated that they felt they acted as an instructor; 9 as a trainer; and 11 as a coach. These data offer hope for the development of IT capability.

APPROACHES TO CHANGE

Why should teachers change? Change is likely to succeed when it lets one achieve goals which are hard to achieve in other ways, it makes life more fun, it makes life easier, and when it is seen to be desirable by the community at large. Many failures to innovate can be predicted in advance, because they fail on too many of these criteria. These criteria contain an implicit agenda for change. Early stages should focus on achieving desirable goals which are hard to reach in other ways (such as supporting problem solving in groups, developing writing skills, inculcating modelling skills), while providing positive classroom experiences for teachers and students. These stages should be attainable without a massive investment of time by the teacher. They should lead onto pleasant classroom experiences, which actually make teaching easier, and the educational gains by students should be judged to be valuable by the students themselves, by other teachers, by parents, and by people outside the school community.

Innovations are likely to fail when they challenge fundamental values and practices, are associated with over ambitious claims by advocates, underestimate the practical constraints of resources such as time and support, are based on exemplary practices, not typical ones, ignore the starting point of the individuals involved, fail to monitor progress, and adapt the programme appropriately. We return to these themes later.

How can institutions change? In our work, we have used survey data to elicit a set of issues which schools commonly identify as barriers to change, and have then used advice from advisers and advisory teachers, the existing literature, our own insights, and evidence from case histories, to show how barriers can be overcome. Of course, the reward from overcoming any barrier to the integration of IT in education is not to succeed - rather it is to face a more interesting and challenging barrier - so lack of hardware is replaced by the need for staff development, which creates a demand for more and better hardware ... and so on. An important outcome from our analyses is a set of ideas on developmental stages along a number of dimensions, which can be used both to describe the current state of development of any school, and to offer advice on what the next stage of development to be tackled is likely to be. The analytic framework is flexible, yet it leads to specific advice for action. Investigative tools are well developed; strategies for progress are grounded in case studies of things that have worked well, elsewhere, yet the framework is not prescriptive - the order in which major barriers are tackled is not pre specified, and a number of routes can be taken to circumvent any given barrier. One way to conceptualise change is offered in Figure 1 opposite.

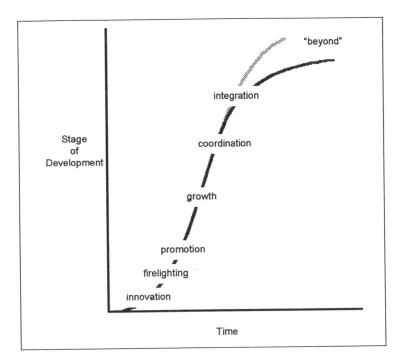

Figure 1: Observable stages of school IT development, adapted from Passey and Ridgway [13]

At a school level, our observations show that the integration of IT into whole school practice takes a minimum of 3 years, and that schools go through at least seven discernible stages in reaching this position (taken from [12]):

- innovation (when one person is finding out what IT is and how it can be used);
- firelighting (when that person is trying to persuade influential people in school about using IT);
- promotion (when school management actively supports the development via INSET, and extra resources including time)
- growth (when teachers begin to use IT more widely, and demand can rapidly outstrip the available resources);
- co-ordination (when the need to monitor students' total experiences becomes urgent);
- integration (when most teachers use IT, a stable state is reached, and IT use is planned and co-ordinated); and
- extension (when new educational uses for IT are explored, and built into students' everyday experiences).

An important insight is that different stages concern teachers with different levels of expertise and autonomy. Initially, development can be entrusted to autonomous enthusiasts; later far more control and central planning is needed. It follows that quite different development needs are met at each stage of development.

A matrix of development produced for the NCC [14] shows the factors which STAC has identified as being important to institutional IT development at each stage. Schools can use this matrix to identify where they are, the factors needing to be considered next, and ways to plan activities to reach their long term goals.

A sample from the entire matrix is given in Table 6 opposite. The sample has been chosen to highlight aspects of school development which relate directly to teacher development.

Our observational evidence (from one day visits to some 24 schools per year), and that from Stephen Steadman [15] indicates that most schools are at the stage of 'Promotion'. Many co-ordinators at the ACITT Conference in Bath (July, 1993) reported that their schools were at the stage of 'Growth'.

Given the current state of development of most schools, and from the time scales which our schools are prepared to consider seriously, we believe that the majority of UK schools will take between 3 and 7 years to reach the stage of 'Integration'. Reaching this point will be completely dependent upon the direct support of senior management in school, the continued commitment at the national level to a strong role for IT in the curriculum, and to the provision of extra physical resources and support. Our model to support systemic change is set out in the chart opposite.

CONCLUSION

The STAC approach to school change is as follows:

- find out where the school is, in terms of teachers' knowledge, current practices, and willingness to use IT;
- establish reasonable constraints:
     - what physical resources could be made available, and sustained in the long term;
     - how many teachers, and in which departments, are willing to become more involved with IT in their teaching;
     - what training could be undertaken, in what time scale;
- establish a range of possible future developments at departmental level, which all fit these constraints;

Factors:	Stages of Development		
	Innovation	**Promotion**	**Co-ordination**
IT is a priority for	individuals	supportive managers	the majority of teachers
teacher awareness of IT use	individuals	increasing numbers of teachers	the majority of teachers
focus of development	self motivated individuals	departmental groups	whole school
roles and responsibilities of staff for IT use and development	defined by individuals and seen by the majority as not their problem	spread widely, but not necessarily understood or universally agreed	agreed and understood by all
forms of staff development	self determined	awareness raising	sharing classroom practice
teaching styles employed in using IT in lessons	specified by one IT expert teacher	explored by small numbers of teachers	wide variety, known about within and across departments
departmental deployment	IT is a specific subject department	each department finds advantageous uses of IT	school development is planned according to department willingness and readiness
student awareness and IT use	restricted to particular lessons	increasing use of IT in lessons outside specialist ones	use of IT in all lessons is co-ordinated
assessment and recording of IT capability	done by one person, if at all	some information about students shared across staff	schemes are in place; teachers understand what is to be learned, and how this can be recorded

Table 6: Examples of factors which influence the integration of IT at different stages of development (Passey & Ridgway, 1993)

- look at the curriculum impact, from the students' viewpoint, of rival possible future developments;
- make recommendations on the basis of the likely coherence and progression of student experiences;
- offer suggestions on how developments can be monitored as they progress.

A range of tools are available to support this process [13]. More are needed. A large number of case histories of schools attempting to integrate IT are necessary in order to build up a broad conceptual understanding of the nature of IT-related change. The uses of IT in education are at a rudimentary stage; the long term potential of IT to shape our conceptions of the nature of learning and knowledge, and the ways that education is conducted, is largely unknown. We believe that the most powerful agents of change are classroom teachers, and that their concerns, beliefs and competencies must be a central focus for professional development.

REFERENCES

1. Department of Education and Science and the Welsh Office (1991) *Science in the National Curriculum.* HMSO: London

2. Department of Education and Science and the Welsh Office (1990) *English in the National Curriculum.* HMSO: London

3. Department of Education and Science and the Welsh Office (1991) *Mathematics in the National Curriculum.* HMSO: London

4. Department of Education and Science and the Welsh Office (1990) *Technology in the National Curriculum.* HMSO: London

5. Passey, D. (1993) Cumbria Education Service Curriculum Information Technology Advisory Centre (CITAC): *Final Evaluation of the Project, 1988-1993.* STAC: Lancaster

6. Gower, S. (1992) *Primary School Teachers and Computers - A Survey.* Undergraduate Project, Department of Psychology, University of Lancaster: Lancaster

7. Passey, D. and Ridgway, J. (1992) School Consultation Report No. 1 *Report of Staff Interviews and Documentary Evidence, Recommendations and Suggested Actions.* STAC: Lancaster

8. Passey, D. and Ridgway, J. (1992) School Consultation Report No. 2 *Staff Development and Planning in IT: Report on findings from a staff survey, with suggestions and recommendations for action.* STAC: Lancaster

9. Passey, D. (1994) School Consultation Report No. 3 *A Report arising from Staff Interviews and Documentary Evidence which offers a Strategic Institutional Development Plan for Information Technology.* STAC: Lancaster

10. Department for Education (1993) Statistical Bulletin 6/93: *Survey of Information Technology in Schools.* Analytical Services Branch, DfE: Darlington

11. Watson, D. (ed.) (1993) *The ImpacT Report - An evaluation of the impact of Information Technology on children's achievements in primary and secondary schools.* Department for Education and King's College London Centre for Educational Studies: London

12. Passey, D. and Ridgway, J. (1994) *The Current Impact of IT and the Stages of School IT Development: Are There Any Prospects for the Future?* Computer Education,   **76** 2-5

13. Passey, D. and Ridgway, J. (1992) *Coordinating National Curriculum Information Technology: Strategies for Whole School Development.* Framework Press: Lancaster

14. Passey, D. and Ridgway, J. (1993, In Press) *Building Information Technology into the School Development Plan.* NCC: York

15. Steadman, S. (1993) *Personal Communication.* University of Sussex: Brighton

Jim Ridgway is Reader in the Department of Psychology at Lancaster University. He has conducted research on informatics over the past decade, and his interests range from analyses of intellectual transactions when computers are used, through to the description of systems dynamics which support or hinder the development of informatics at a national level. Currently he is director of the STAC Project which produces a range of materials to support school uses of computers. He also has major interest in issues related to assessment.

Don Passey is Senior Research Fellow in the Department of Psychology at Lancaster University with the STAC Project. He has carried out research into educational change related to use with informatics, how support can be provided for those developing use with informatics, and how co-ordination of uses of informatics can be managed within educational institutions. He has authored and co-authored a range of publications in these areas, and has provided consultancy to schools in the UK and abroad.

7

The reality of learners' achievements with IT in the classroom

David C. Johnson

Shell Professor of Mathematics Education
King's College, University of London, United Kingdom

ABSTRACT

Is there a link between surveys of resources and use and the research on pupil's achievements using IT? Although evidence of an increasing level of provision, and reports on learning with IT indicate the potential, there is still the question of the reality of classrooms. The UK ImpacT research indicates that within 'computer using' classrooms the benefits can be positive. However, these benefits are dependent upon the nature of the experiences and aspects of teaching style and classroom management. Even with a high level of commitment to the acquisition of equipment, the present situation is still deficient in terms of pupil and teacher access and teacher support.

Keywords: information technology, learning, research, teaching methods

INTRODUCTION

".. notebook computers are producing positive results in the school science laboratory." (Times Educational Supplement, "IT Review", 7 January, 1994, p. 26)

"... this month's (UK) technology in education (conference) clearly demonstrated how quickly IT has become part of the fabric of most schools and colleges." (Times Educational Supplement, "Computers/IT", 21 January, 1994, p.17)

The first editorial above goes on to note that students were "confidently using the data-logging equipment", and further that word processing and spreadsheets were also "proving valuable". And in the second, a number of recent commercial developments were described, along with statements suggesting, for example, a "dramatic growth of CD-ROM in education" and exciting developments in the use of interactive optical disc systems. Very little, however, was said about the actual activities or achievements of the learners, their expectations and behaviour in these integrated informatics environments, nor where one might readily observe this phenomenon. While it does seem clear that developments in hardware and software offer truly exciting potential, what do we know about the reality of classrooms? Is this integration really as common-place as suggested in the press? Further, does the 'integration', if it is really taking place on any large scale, provide benefits beyond training in the use of the technology?

There exists considerable evidence related to the theme of learners in an integrated IT environment, which has been generated through a range of data collection activities. One main source are the large scale surveys of availability of resources, for example numbers of machines and pupil/machine ratios, and aspects of access and use in school subjects, often presented as percentages of total computer use distributed across subjects such as maths, science, native language, etc.. In the UK such work has been carried out on a regular basis by the Department for Education (see [1], [2], and [3]). At the international level we have the multi-national IEA Computers in Education Study [4], and, of particular relevance to this paper, Anderson's [5a] report on the US component of this work. Anderson reports on a large scale survey (over 69,000 pupils, aged 10, 13, and 16, in 2,500 schools) conducted in 1992, a follow-up to a similar study conducted in 1989. The report provides information on percentages of teachers in "computer-using schools" who use computers in class "on at least several occasions during the year", typically around 50%. The schools also had a pupil/computer ratio of about 7/1, and note is made of the fact that since 1989, U.S. schools have increased their inventory of computer units by as much as 50%. In making international comparisons, Anderson states that "in ... peripherals and networks, the US has at the very least comparable and often larger inventories than the other countries (in the IEA survey)" p. 27.

The US study also placed emphasis in the research upon using computers in learning different subjects such as science, geography, languages. "In the U.S., about half of school computer time is spent on learning about computers and the other half on learning other subjects with the help of computers" (p. xvii). But what does this "learning other subjects" really mean? Some aspects of this question are discussed by Lundmark in the chapter "pupils' opportunity to learn with computers" [5b], but the evidence reported is primarily that of "counts" of responses to items in the survey questionnaire as opposed to actual observations of classes and pupil assessments. An important consideration in the discussion is the fact that pupils were counted as a "substantial user" of IT, if they used software six or more times in the school year, i.e. six or more times in a period of 150 school days. In her conclusions, Lundmark notes that:

> "The question of how far computer instruction has been integrated across the school subjects curriculum is noteworthy. A little more than one-fourth of all US students say they used computers in none of the traditional subjects (included in the survey) and a little more than one-fourth say they used computers in only one of those ... subjects during the year. ... it seems that the use of computers as instructional tools has not advanced very far across the curriculum." p. 70.

A further source of evidence for discussing learners and IT in school subjects, are the numerous studies conducted in different educational contexts, for example, multimedia learning environments or aspects of group work in classrooms, or studies which focus on the effects of IT on pupils' learning within a specific conceptual domain in a curriculum area. Niemiec and Walburg [6] report on results from over 250 studies on the effects of IT on pupils' learning of particular skills and concepts, with the overall results generally supportive of the experience.

Evidence can also be found in books, monographs and conference proceedings, which summarise and synthesize the research and theoretical perspective supporting classroom activity linked to a particular type of use such as simulations, or the role of powerful generic software tools [7, 8]. Conference proceedings often include sub-themes which provide some degree of synthesis, either within or across some classification [9].

As already indicated, the evidence from these, and other sources (e.g., the practical materials produced for teachers) is generally supportive of the role of IT. Some research evidence links the surveys of access and use with individual studies with their focus on classroom activity in a well-defined subject context; other reports synthesize work in support of a particular type of activity addressing issues of teaching and learning in the wider context of 'real' classrooms. Such evidence is, at best, limited. What can we say about the impact of IT on teachers and pupils activities and achievements, across subjects and over time in the world of "typical" schools? Are results only possible in restricted and highly supported (both in terms of people and resources) contexts, as typified by those described in many of the individual studies reported in the literature? Is it still a case of "we know what might be achieved, but are not in a position to reach this potential in most schools?"

THE IMPACT RESEARCH (ENGLAND AND WALES)

The UK ImpacT study [10a]; "an evaluation of the impact of information technology (IT) on children's achievements in primary and secondary schools", was carried out in the period January 1989 - December 1991. While status data dealing with resourcing and use comparable to some aspects of the US study were collected, a main focus of the research was on pupils' achievements and classroom activity involving IT. Assessments and monitoring were conducted in the subject areas of mathematics, science, geography, and English, over a period of two years with pupils in the age ranges 8-10, 12-14, and 14-16. The sample involved over 2300 pupils from 87 classes in 19 Local Education Authorities in England and Wales.

The ImpacT work both extends and links aspects of the research literature base in this area to include longitudinal effects within school subjects, cross subject considerations of general aspects of classroom use of IT, and the provision and use of hardware and software resources. These were integrated to address a range of issues encompassing learning, pedagogy, and policies. The ImpacT results from the component parts were linked to enable the research team to address three main questions:

- Pupils' Learning: did IT make a contribution?
- Pedagogy and Practice: what can we say about the planning and practice of teaching to incorporate IT?
- Schools' Organization: what were the demands of IT on the schools?

The focus of the remaining portion of this paper is on the first question, which represents one main component in considering learner's outcomes, within an integrated informatics environment, with selected findings related to the second question used to inform or extend some aspects of the discussion.

Methodology
One important feature in the ImpacT research design was the inclusion of pupils from matched pairs of classes. One of each pair was identified as

making regular use of IT, designated HiIT where the teacher was an experienced user of IT and had plans for regular use during the period of the research. The initial confirmation of this designation was a non-trivial undertaking. The HiIT class was then matched with another class for which the teacher had no plans for using IT, making low, or no, use of IT, and thus designated LoIT. LoIT class teachers in particular were also nominated for their good teaching and curriculum delivery; it was not merely a case of finding volunteers. The pupils divided into a matrix of 12 cells, the three age groups crossed with the four curriculum areas. The sample was used in a framework for data collection which had three substantial parts:

• An assessment of pupils' achievement, through the administration of specially designed subject focussed assessments. These were supplemented by eight topic specific mini-studies in some pairs and also some HiIT only classes, and a final test for IT concepts and skills. Statistical comparisons of test performance were adjusted through the use of general ability assessments.
• Five in-depth longitudinal case studies in HiIT classes, focussed on classroom processes and pupil interactions. Classrooms were observed, pupils and teachers were interviewed and documentary evidence was gathered to provide illumination on classroom realities. Qualitative analysis was based on those themes and issues that emerged from the data.
• IT resourcing and use was monitored throughout by the regular returns of questionnaires and data sheets from the teachers and pupils in each class. Hardware and software provision, pupils' IT use in ImpacT subject and across all subjects, and pupils' extra-mural use were analysed descriptively by classes, age cohorts and school subjects.

Further details on the research design and methodology are given in the full report [10b] and a recent paper [11].

ImpacT: Results and discussion
Within the limitations of the study [10c], the answer to the question of "the contribution of IT use to learning" was yes, IT did make a difference, but the contribution was not consistent across school subjects or age bands. The global results in support of this response were from the main field study subject reasoning assessments; post-tests administered twice for each subject. These were also supported by the results of the 'topic specific' mini-studies. An overall summary of the HiIT and LoIT group outcomes, subject reasoning and mini-studies, is given in the table below.

In the case of subject reasoning, these represent combined effects for both assessments in each subject age band.

Subject	Mathematics		Science		Geography		English	
Age	Subj. Reas.	Mini Study	Subj. Reas.	Mini Study	Subj. Reas.	Mini Study	Subj. Reas.	Mini Study
8-10	**pos.**[a]	no study	**neg.**[b]	incon.[c]	age group was not included		**pos.**[d]	**pos.**
12-14	incon.	**pos.**	incon.	no study	incon.	no study	incon.	**pos.**[e] incon.[e]
14-16	**pos.**	no study	incon.	**pos.**[e] incon.[e]	**pos.**	**pos.**	insufficient data for analysis	
a. **pos.** - positive result, favoured HiIT. **b.** neg. - negative result, favoured LoIT. **c.** incon. - inconclusive or no differences. **d.** positive result was partial **e.** two mini-studies were conducted with this subject/age group.								

Table 1: Summary of ImpacT Results: Field Study Reasoning-in-Subject Tests and Mini-studies

The major focus of the subject reasoning assessments was on higher order processes and thinking deemed appropriate to each subject; for example, relational thinking in mathematics; formulating hypotheses and designing experiments in science; drawing inferences from map, graphical and photographic information in geography; and aspects of content and cohesion/coherence in pupils' writing in English.

When these results were probed at the level of classes it was the case that the results from a small number of the HiIT classes provided the main evidence for the findings reported in Table 1. Access and use in these classes suggested there may well be some minimum threshold of IT use for the impact to be detected. The results from the US survey and the ImpacT resources and use data, suggest such access is not a common phenomena.

In terms of topic specific learning outcomes, five of the eight mini-studies provided evidence in support of the affirmation that IT use contributed to learning. This is consistent with the results from 'topic specific' research reported in the literature. However, in the case of the ImpacT research, the results also indicated that the contribution was in terms of higher level processes or thinking. For example, in the mathematics mini-study on angles the two classes were studying the same materials, with the pupils in the HiIT class also working with the programming language Logo. The HiIT class achieved significantly higher results, with the main contribution being to the higher level processes, i.e. application of the concepts and relationships [12]. Similar contributions were noted for:

- a HiIT-LoIT science mini-study dealing with relationships between elements where the HiIT class used a database of chemical elements;
- a HiIT only geography mini-study 'indicators of development' in which the pupils were observed while using the world countries database PCGlobe;
- two mini-studies in English, which involved aspects of pupils' writing using word processing (one a HiIT-LoIT comparison and the second HiIT only).

While one of the science mini-studies indicated a contribution from the use of IT, this was offset by the failure to detect any differences in the other two mini-studies and the consistent negative or inconclusive results for the science subject reasoning assessments. However, it was also the case that science represented the most difficult area for the identification of HiIT classes. Computer use in science was generally low when compared with the use in the other ImpacT HiIT classes. While the relative percentages of IT use at the international level also support this finding, it still remains a paradox in that much work has been done in hardware and software development to support learning in science; how does one explain this apparent contradiction?

As indicated in the research methodology, the main focus of the case study research was on classroom processes. While the data collection was rigourous and detailed, the analyses were designed to provide exemplification rather than generalisations, and as such serve to illuminate rather than explain the achievement outcomes. Selected observations from the case study classes suggest some important considerations associated with pupils' use of IT:

- computers were found to be good motivators which heightened pupils' interest and enjoyment and were also seen to have a positive effect upon the status of the subject;
- computers aided concentration by focussing pupils' attention on the work in hand and as a result some pupils and teachers believed that the standard of work produced was of a higher quality than it would have been otherwise;
- opportunities to work in an open-ended way enabled pupils to become involved in more complex and challenging learning situations beyond that typically experienced.

Further, some of the failures to detect any contributions through the use of IT may be attributed to some of the problems encountered by pupils in the case study classes:

- difficulties in using a particular software package;
- inability to work effectively in a collaborative environment.

The link of achievement results to teacher style or planning provided some further insights in terms of pedagogy and practice:

- The answer to the question regarding the planning and practice of teachers mainly involved a consideration of classroom management and organisation and teaching styles along with aspects of hardware and software availability and use. The results from the case studies, mini-studies, and aspects of the field study, indicated quite clearly that any contribution was dependent upon a range of factors, the most important being that of the role of the teacher.
- Teachers' responsibilities were found to demand careful attention to organisation and management, in particular the effective use of collaborative or group work. Further, effective use of IT represented substantial demands in terms of knowledge and understanding of, and familiarisation with, a variety of software in order to integrate the activity, in philosophical and pedagogical terms, within a larger scheme of work.
- The use of more general purpose software, for example spreadsheets, databases, and programming, placed additional demands on the teacher, beyond that of becoming familiar with the use of the more complex software, to include more reflection on the nature of their subject and the potential role of such software in enhancing processes and understanding of 'big' ideas or relationships.

These points are quite consistent with reports on research cited earlier in this paper, in particular those works which attempt to synthesize results across a wider domain of studies. On the other hand, if we are truly working towards an integrated informatics learning environment for all, the integration of findings and the linking of such evidence to the world of 'real' classrooms demands attention to a multitude of issues. Two issues of importance here are those of access and teacher support.

CONCLUSIONS

Research, including that from the ImpacT study described above, has confirmed the potential for significant contributions of IT to pupils' learning. It is proposed here that any consideration of the notion of "learner's expectations and behaviours in an integrated informatics environment" is of course firstly dependent upon planned experiences being available through the provision of both resources and opportunities to access and use the resources to engage in challenging tasks. As indicated above, the role of teachers is crucial to this endeavour, and this in turn is dependent on their perceptions as to the benefits from such opportunities and their ability to take action. Of particular importance are their abilities linked to teaching styles and classroom management, along with, of course, those widely acknowledged capabilities of rapport, insight, and capacity for work. Attention is directed here to two recent insightful reports on case studies of teachers who have effectively integrated computing into their instructional processes [10d], [13].

While the role of the teacher is accepted as of paramount importance, it is also the case that all involved, teachers and pupils, need "ready" access to equipment. What does this really mean? It is accepted that government, local education authorities and schools have made efforts towards increasing the numbers of machines available to pupils (decreasing the pupil/machine ratio), and, for many countries, there have been some dramatic changes. However, while Anderson recommends a ratio of 5/1 by 1995 [5a], and such might seem encouraging, I would suggest that even this is far from adequate. A recent computer manufacturer newsletter, Acorn *Arc*, Autumn 1993, notes a goal of "one micro per child" p.18, but my own suggestion is that *"schools, as business, should have more computers than people"*.

This is not to say we need fewer teachers, but rather that the resources available should be such that teachers and pupils each have their own computer, and the provision of other resources, such as electronic communication, multi-media or CD-ROM and interactive optical disk, should be of an order to facilitate the full range of potential activities in teaching and learning supported with IT. The goal of maximising the use of available equipment, so that machines should not be idle during the day, for reasons of cost, is inappropriate in contexts which look to optimising the productivity of people.

With this in mind, I end this paper with the question; "How would you feel if you had to book time on your machine?" I would be truly lost without the 'Apple on my desk' to use when, and if, I feel the need. The idea that I should share or book time, because the technology is not being used throughout the day, would represent a real intrusion. What does this say about "learner's expectations and behaviours"?

REFERENCES

1. D.E.S. (1986) *Results of the Survey of Microcomputers in Schools.* Statistical Bulletin **18/86**. London: D.E.S.

2. D.E.S. (1989) *Results of the Survey of Information Technology in Schools.* Statistical Bulletin **10/89**. London: D.E.S.

3. D.E.S. (1991) *Results of the Survey of Information Technology in Schools.* Statistical Bulletin **11/91**. London: D.E.S.

4. Pelgrum, W.J., and Plomp, T. (1991) *The Use of Computers in Education Worldwide.* Results from the IEA Computers in Education Survey in 19 Education systems. Oxford (England): Pergamon Press.

5a, b. Anderson, R.E. (Ed.) (1993) *Computers in American Schools 1992: An Overview.* A National Report from the International IEA Computers in Education Study. Minneapolis (Minnesota): University of Minnesota (Social Sciences).

5b. Lundmark, V.A., *Opportunity to Learn with Computers* Chapter 5, 55-70.

6. Niemiec, R.P., and Walburg, H.J. (1992) *The Effects of Computers on Learning.* International Journal of Educational Research **17**(1).

7. Hoyles, C., and Noss, R. (Eds.) (1992) *Learning Mathematics and Logo.* Cambridge (Massachusetts): The MIT Press. Note the focus on the learner in the papers by Kynigos, *The Turtle Metaphor as a Tool for children's Geometry*, 97-126, and Hoyles and Noss, *Looking Back and Looking Forward*, 431-468.

8. Schwartz, J.L., Yerushalmy, M., and Wilson, B. (Eds.) (1993) *The Geometric Supposer: What is it a case of?* Hillsdale (New Jersey): Lawrence Erlbaum Associates, Publishers. Note the focus on the classroom in the chapter by Wilson, 17-22, and the ideas discussed regarding the problems of teaching in Part III (notions of promoting pupil 'enquiry' and 'thinking').

9. Johnson, D.C. and Samways, B. (Eds.) (1993) *Informatics and Changes in Learning.* Proceedings of the IFIP TC3/WG3.1/WG3.5 Open Conference, Gmunden, Austria, June 1993. Amsterdam: North Holland.

10a,b,c,d. Watson, D.M. (Ed.), (1993) *The ImpacT Report: An evaluation of the impact of information technology on children's achievements in primary and secondary schools.* London: King's College London (for the Department for Education).
10b. Cox, M.J. *The Project Design and Method,* Chapter 2, 7-25.

10c. Johnson, D.C. *Summary, discussion and implications,* Chapter 7, 153-166.

10d. Watson, D.M., Moore, A., and Rhodes, V. *Case Studies,* Chapter 4, 61-96.

11. Johnson, D.C., Cox, M.J., and Watson, D.M. (1994) *Evaluating the Impact of Pupils Use of IT on Learning.* Journal of Computer Assisted Learning **10** (3).

12. Johnson, D.C. (1994, in press) *Logo and Angles: A Focussed Investigation with Pupils Aged 12-13.* In Y.Y. Katz (Ed.) Computers in Education: Pedagogical and Psychological Implications. (IFIP).

13. Kearsley, G., Hunter, B. and Furlong, J. (1992). *We Teach with Technology: New Visions for Education.* Wilsonville (Oregon): Franklin, Beedle & Associates.

David C Johnson took up the post of Shell Professor of Mathematics Education in the University of London in 1978, following 17 years at the University of Minnesota (US). His teaching experience spans all levels from primary through university and he has been involved in school computing since 1963; activities which utilised hardware starting with mainframes in the early days, through time-share and leading on to today's micros and powerful new technologies, software and supporting equipment. He has been a member of WG3.1 (and contributor to the working group publications) since its early days

8

A Curriculum for teachers or for learning?

Paul Nicholson

Deakin University, Australia

ABSTRACT

What should the focus of an informatics curriculum be? The majority of existing courses mimic industrial or commercial practice and reflect a particular vision of teaching and learning. Humanistic and constructivist perspective's suggest that alternative course models can empower students to become independent learners capable of operating in the complexities of modern society. The implementation of such a curriculum focus would place enormous strains on teachers and schools. Is this approach worth pursuing? What are the advantages to learners and what are the costs to schools?

Keywords: curriculum policies, information technology, learner centred learning, pedagogy, philosophy

INTRODUCTION

As we approach the middle of the third decade of widespread computer use in schools, it is timely to reflect on the evolution of classroom pedagogy and curriculum practices and to review how the computer has changed the curriculum, the nature of teaching and learning and indeed schools themselves.

In the first decade, loosely 1970-80, technocentric paradigms [1] of computer use drove the curriculum and dominated classroom practice.

Limited hardware and software capabilities and transmissive models of teaching and learning dominated practice and much was expected of "drill and practice" software, particularly in mathematics. Computers were expected to redefine the role of the teacher and to bring about new efficiencies in schools. By the end of the decade, concerns were being raised about the lack of real improvement in the nature and quality of classroom teaching under these paradigms. Learner centred paradigms began to emerge [2], signalling the development of broader under-standings of how computers might be used to facilitate learning across a range of disciplines. Concurrently, educators began to plan for the wide scale use of computers across the curriculum [3].

"Computers across the curriculum" may well have been the catch cry of the 1980's (the second decade) with computers being applied to most discipline areas. In this decade, a diverse range of paradigms of use existed concurrently. Many of these reflected technological improvements in hardware or software developments, but were still essentially based on the older technocentric paradigms. The emergence of "Information Technology" (IT) as an identifiable entity had a profound effect upon the curriculum with the implementation of courses related directly to business and commerce, or aligned to national imperatives. The imposed need to develop IT skills affected the curriculum much as the Sputnik affected American science education. Towards the end of the decade, the nature and role of IT was being seriously questioned, and educators began to adopt the learner-centred paradigms alluded to a decade before [4-8].

In the 1990's, computers are in wide-scale use in schools in most technologically advanced nations. The rationale for their ongoing place in the curriculum, their mode of use and their effectiveness in achieving their perceived goals forms the basis of the remainder of this paper.

WHAT KIND OF CURRICULUM?

What kind of informatics curriculum should we provide for our students? What vision of society, individuals and learning should it embody? Why? These key questions are often overlooked in the pragmatics of planning or imposing curriculum on schools and school systems, especially in the areas of computing and technology. I believe that the issue revolves around the perspective that schools adopt about what a curriculum should *do*, and for whom. At one level many curriculum projects, for example the Australian National Curriculum project (in development), are about defining and classifying the scope of activities that students are exposed to. They are essentially planning documents for teachers and schools to use to ensure that their curriculum has satisfactory "breadth". While it includes information about student profiles (performance indicators), it is essentially an organisational model of curriculum development not all that different from those in the UK. In this paper I do not wish to debate the nature and content of such imposed curricula, even given their importance to schools, but rather to focus on two diametrically opposed curriculum models which could be (and are) used by schools in implementing an informatics curriculum.

Technocentric curriculum models

For most of this century, the dominant educational paradigm of western countries has been a technocentric one, delivering education in a Taylor-like mass production process [9]. This is particularly true in the classroom use of computers where there is a new emphasis on learning about technology, that is, computers, to be able to better meet the perceived needs of industry. In these models, the purpose of the information technology curriculum is typically to provide:

- some knowledge and understanding of computers and their technology;
- some knowledge and understanding of the impact of computers on society;
- some knowledge of computer applications and the use of a variety of contexts for them;
- the ability to use a few standard types of software;
- the ability to write some simple programs.

This type of curriculum is usually presented as a senior IT course, or at a lower level, as "computer awareness". In many cases there is little attempt to integrate the use of this computer knowledge into the mainstream curriculum; the computer remains a technological entity to be studied in an abstract way; interesting, relevant, but not in my subject. The current UK IT curriculum is a good example of this type of curriculum. Papert believes that the localisation of computers, that is, withdrawing them from classrooms into laboratories and the formalisation of their use in IT courses, is *the* major curriculum issue of this decade [10].

The relevance of these pragmatic curriculum models, and their frequently associated behaviourist models of teaching, learning, and assessment to the challenges of modern society are being increasingly questioned by academics and philosophers [11-13] who see alternative ways for individuals to participate in a society heavily reliant on information technology and rapid access to information.

The institutional response to criticisms of the traditional (technocentric) informatics curricula has been to develop more flexible instructional systems. A typical example is the "Ford Academy of Manufacturing Sciences" program which aims to provide students with a wider range of educational experiences within a business context. The focus however is still instructional and essentially reductionist; focusing on the future needs of industry by ensuring that students gain a wider range of specific business-related skills than they would have had previously.

Currently, many national educational policies implicitly or explicitly locate schools within what is now termed an "economic-rationalist" model in which schooling serves as preparation for participation in the industrial and economic spheres of society. The commercialisation and corporatisation of education by business is a major concern as schools increasingly struggle with the dichotomy between egalitarian models of education and funding sources based on sponsorship models and targeted marketing policies [14]. Implicit in this business relationship is the idea that schools will in some way embrace the goods, services or philosophy of the sponsor or provider. For example in the United States, both Apple Computer and IBM have initiated massive sponsorship of educational programs, ranging from single schools to whole school districts; for example the Electronic School District Project in Indiana [15]. In the overall context of American education, these sponsorships are trivial in that the system is large and robust enough to absorb them as interesting experiments or case studies. They do not, of themselves, become the mainstream however much the sponsor would like them to. In smaller or less well resourced countries, this is a real danger and it must be guarded against. Hawkridge's analysis [16] of the short-term political benefits to third-world countries of implementing "computers in education" should be kept firmly in mind here as corporate sponsorship usually provides ready access to technology without providing an underpinning pedagogical basis for its classroom use. There is also a distinct lack of concern, regardless of marketing rhetoric, with the quality of education that results, and often a desire to measure learning outcomes in a simplistic way that ignores the complexities of human learning.

The emergence of post-structuralist, post-modern curriculum theories and the demise of positivist paradigms provides a set of theoretical, analytical frameworks with which to deconstruct these older curriculum models. However whilst informing us about the real agenda behind these curricula, they do little to inform us about what might replace them. Fortunately, the last decade has seen an increased interest in the psychological foundations of learning, and of ways of using computational media to facilitate learning not teaching. The research in this area provides many useful guidelines about ways of using computers in schools that are different from the traditional IT focus, concentrating instead on ways to bring about quality learning.   To implement these ideas though will require a major change to the nature and content of the school curriculum and classroom practice.   The fundamental change is a shift from technocentric thinking about "using computers in the curriculum" to humanistic thinking about the same issues.   The change is essentially a shift from "learning about the machine" to "using the machine in learning".

A humanistic curriculum model
"The computer should be like a pencil, not as an isolated class, but as a tool which empowers children with knowledge, thinking skills and problem solving alternatives" [17]. This is the fundamental tenet of humanistic computing.   It represents an approach to education and learning which is essentially constructivist (or constructionist, in Papert's terminology) in nature and which is based on a substantial body of educational practice and psychological research. The constructivist model of learning is radically different from the more traditional models that predominate in schools. It has a focus on learning rather than on the use of technology.   It provides schools with a key issue to discuss in planning their informatics curriculum - what are we trying to do with computers in schools, and why?

A curriculum based on constructivist guidelines should at least have a focus on:

- the student as the constructor of knowledge;
- independent and flexible pathways through traditional curriculum areas;
- an appreciation of the difficulties of "realistic assessment" as currently being debated in the literature;
- the use of computational media to facilitate learning.

While these sound like utopian items, and for many schools in many countries they are, there are a few schools actively involved in attempting to implement some or all of these items in their curriculum. The problem that they face is in identifying a suitable curriculum model. Perhaps the most widely understood, and best researched in this paradigm, is Papert's computational model [18].   In this model, computers act as personal knowledge tools in an environment in which students can engage with, and explore, their developing knowledge, using Logo as a programming tool. Some research has shown that Logo programming can produce marked conceptual change and alter the role of teacher and student in the construction and transmission of knowledge [19]. However, while Papert can display successful examples of his philosophy being implemented in schools, these generally occur only in special environments. For example both the Henigan school in Boston and Melbourne's Methodist Ladies' College (MLC) are both atypical schools in that the former is heavily supported by MIT and the latter by a large investment of private school funds. A key to MLC's successful adoption of a "Logo-like" curriculum focus has been an extensive, longitudinal program of staff development and assistance in the purchase of computers by staff. Additionally, MLC has shifted much of the cost of the innovation, approximately $2,000 (AUD) per child, onto parents of new students, who are required to have access to an IBM compatible (386-SX or better) lap-top computer from grade 7 onwards. The junior school (pre grade seven) also uses lap-top computers in its curriculum and some parents have opted to buy them for their children's use, whilst others have leased them from the school. Other Australian schools have now adopted lap-top computers in their curriculum, but MLC is unique amongst these in having an informatics curriculum solidly based on a defined area of educational research.

In such a focused and well resourced school environment, it could be expected that the benefits of a Papert's (computational) environment would be obvious; with many of the students displaying significant programming skills and being able to apply them to all curriculum areas. While many MLC students can, and must, be advanced Logo programmers, much of their computer use is for word processing and there appears to be a reduction in emphasis on programming for knowledge construction in senior years. In many ways it appears as if the whole concept has become institutionalised, losing its fundamental "student as learner" focus to the incorporation of Logo into a more formally structured curriculum.

Can this essential focus be sustained in "normal" classrooms and average schools? The MLC experience has clearly shown the need for a huge investment in technology, in infrastructure and in staff development. This is not likely to be possible for an average Australian, American or British school system. Does this mean that Papert's model is unrealistic and that such an emphasis on individual computational skills is inappropriate? Can the model be successfully scaled down to individuals in individual classrooms? These are questions which deserve much greater attention from researchers.

CONCLUSIONS

When telephones were first placed in factories, it was expected that the managers would use them to call their workers in order to direct their activities. The shock that occurred when the first worker called the employer must have been significant. Schools now have to face up to similar culture shock issues about the educational use of computers. They must make a philosophical decision about the nature and purpose of public education - who is it for and for what purpose? I would argue strongly that the focus must be on providing the individual with the ability to learn outside of the formal education system, and to develop appropriate skills with which to actively engage in knowledge construction. There is so much evidence about the increase in our knowledge base, and its rate of change, that we should accept that to adopt essentially transmissive curriculum models is irresponsible. We need to find ways in which we can help our students to become masters of, not slaves to, this knowledge base. The school curriculum should be the first place that we start with this major task.

Although the comments in the previous section were confined to Papert's constructionist model, similar concerns have been raised in other contexts, for example science education. There is a fundamental problem that schools face in adopting a humanistic-constructivist curriculum model; it overturns the institutional structure and focus. Institutional inertia and the industrial realities of the teaching profession will ensure that schools do not become the educational shop-fronts advocated by Ivan Illich in the 1970's. It is almost certain that they will retain most of their present characteristics for many years.

In attempting to address these and other "quality" issues in the informatics curriculum, it is clear that a humanistic curriculum has much to offer, putting real knowledge and learning in the hands of individuals. The fundamental, challenging question we have to address is, how can we help schools to adopt such an approach in times when economic models of education are becoming more prevalent? Is a set of "guidelines for good practice" adequate? Do we have to document and publicise exemplary practice? Should we ensure that research findings are more widely published and addressed to the school community? Do we need to write educational monographs for teachers and teacher-training, and should this be a focus? Is "constructivism" a suitable theme for a conference? IFIP WG 3.1 has a broadly based membership which should be able to address, at least in part, all of these issues..

REFERENCES

1. Papert, S. (1987) *A Critique of Technocentrism in Thinking About the School of the Future.* Children in an Information Age: Opportunities for Creativity, Innovation and new Activities. Sofia, Bulgaria.

2. Abelson, H. and P. Goldenberg, (1976) *Student Science Training Program in Mathematics, Physics and Computer Science.* Final Report to the National Science Foundation. Massachusetts Institute of Technology.

3. Allenbrand E. (1979) *Course Goals in Computer Education, K-12* Tri-County Goal Development Project, P.O. Box 8723, Portland, OR 97208: Report Number 617.

4. Barta, B., et al. (1990) *Computers in the Israeli Educational System - Implementation Aspects (1984-1989).* in McDougall, A. and Dowling, C (eds) Computers in Education Sydney, Australia: Elsevier Science Publishers.

5. Bosler, U. (1990) *New Information Technology at the Secondary Level: A Survey of the Situation in the Federal Republic of Germany.* in McDougall, A. and Dowling, C (eds) Computers in Education Sydney, Australia: Elsevier Science Publishers.

6. Driver, R. and E. Scanlon (1989) *Computer Experience, Self-Concept and Problem-Solving: The Effects of LOGO on Children's Ideas of Themselves as Learners.* Journal of Educational Computing Research, 1989. **5** (2) 199-212.

7. James, C.B. (1989) *Observations on Electronic Networks: Appropriate Activities for Learning.* Computing Teacher, **16** (8) 17-19.

8. Qi, C. and W. Ben-Zhong. (1990) *Educational Computing in Chinese Schools.* in McDougall, A. and Dowling, C (eds) Computers in Education Sydney, Australia: Elsevier Science Publishers.

9. Feldmann, B. (1993) *Human-Centered Science in Education and Technology.* in Rethinking the Roles of Technology in Education. Cambridge, Massachusetts: The University of Texas at Austin.

10. Papert, S. (1993). The Children's Machine., New York: Basic Books.

11. Dertouzos, M.L. (1989) *Made in America.* Cambridge, Massachusetts: The MIT Press.

12. Naisbitt, J. (1984) *Megatrends.* London: MacDonald & Co.

13. Levin, J.A. et. al, (1989) *Technology's Role in Restructuring Schools.* Electronic Learning, **8** (8) 8-9.

14. Bigum, C. (1994) *Schooling in an age of informed cynicism.* Deakin University: Faculty of Education.

15. Halla, M. (1990) *The Electronic School District Project: Wide Area Networking for Primary and Secondary Education.*in McDougall, A. and Dowling, C (eds) Computers in Education Sydney, Australia: Elsevier Science Publishers.

16. Hawkridge, D. (1990). *Computer's in Third World Secondary Schools: The Ministry's Viewpoint.* in McDougall, A. and Dowling, C (eds) Computers in Education Sydney, Australia: Elsevier Science Publishers.

17. Leventhall, S., P. Stevens, and M. Melton. (1993) *Managing Data Daemons.* in Rethinking the Roles of Technology in Education. Cambridge, Massachusetts: The University of Texas at Austin.

18. Papert, S. (1980) *Mind Storms.* New York: Basic Books.

19. Harel, I. and Papert, S. (1993) *Constructionism.* Norwood, New Jersey: Ablex.

Paul Nicholson studied science at post-graduate level and then taught science and computing in schools, completing his M Ed in 1984. Since 1988 he has co-ordinated post-graduate computer education courses at Deakin University. He is undertaking a PhD in the use of computer-based probes for studying cognition. Paul is the author of several books on physics education, "sysop" of a FIDO bulletin board and conducts training courses in developing multi-media software, instructional design and electronic communications. His research interests are in the study of meta-cognition, the design of multi-media software and the use of telecommunications in education.

9

The case of Cabri-géomètre: learning geometry in a computer based environment

Colette Laborde
Jean-Marie Laborde

Informatics and Applied Mathematics
Grenoble University, France

ABSTRACT

One of the reasons teachers may resist the use of computers in the mathematics classrooms is the change that this introduction implies about the kind of problems fostering the acquisition of knowledge by the pupil. Some old problems become uninteresting in the new environment. Conversely computers allow the design of some new problems that are impossible in a paper and pencil environment. In our paper we consider the computer as part of the "milieu didactique" : as such it offers specific representations of knowledge, specific facilities and specific feedback to the pupils. We will discuss the role of these characteristics on the way mathematical objects may be perceived on the interface by the pupils and on the solving processes of the pupils when they are confronted to problem situations on the computer, using the geometry program Cabri-géomètre.

Keywords: learner centred learning, drawing, graphics, innovation, microworld

INTRODUCTION

Much research provides evidence that listening to the discourse of the teacher, to a clear presentation of mathematical content, does not guarantee the learning which was expected. According to constructivist theory which is widely used among mathematics educators, knowledge is actively built up by the cognizing subject when interacting with mathematical learning environments. This notion of environment must be taken in a broader sense that the environment is of material nature as well as of intellectual nature.

Problems are part of such environments, and may play an important role in the construction of mathematical knowledge by learners because they offer the opportunity of involving the learners' own ideas and of testing their efficiency and validity, when trying to solve the problem. Brousseau [1] considers the process of elaborating a solution by a learner as the interaction between the learner and a "milieu" which enables the learner to perform some given kinds of action in order to solve the problem and offer feedback to actions. In terms of systems, the "milieu" is the system working with, by "replying" to, the learner.

The graphical and computing possibilities of software now allow a reification or restatement of abstract objects and in particular of mathematical objects, as well as numerous possible operations and related feedback. The construction of knowledge may be fostered by problem solving in a computer based environment. But as Pea [2] claimed, the computer provides new tools for operating on these objects and therefore changes the objects themselves. Feedback can also be of a different nature; it may be very sophisticated in comparison with paper and pencil situation. All these points imply that designing a problem in a computer based environment requires a new analysis of mathematical objects, operations and feedback. This explains perhaps why teachers can be reluctant to use computers in their classrooms.

THE NATURE OF GEOMETRY

The nature of geometry is dual in that problems related to geometry may be of practical nature as well as theoretical. Although geometry was originally built as a way of controlling the relations with physical space. It has also been developed as a theoretical field dealing with abstract objects.

One usual mediation of the theoretical objects of geometry is offered by graphical representations called "figures". The relation of the mathematicians with these representations is complex and in a sense reflects the dual nature of geometry; mathematicians drawn them, they act on them as if they were material objects (by means of a kind of experimentation), but their reasoning actually does not deal with them but with theoretical objects. This is why Parzysz [3] distinguishes the "figure" which is the theoretical referent (attached to a given geometrical theory : euclidean geometry, projective geometry...) from the "drawing" which is the material entity. But geometry software enabling the drawing of figures on the screen of computers forces us to refine this distinction in order to account for this new mediation.

From drawing to figure
As a material entity, a drawing is imperfect - the lines have a width, the straight lines are not really straight. But mathematics ignore these imperfections and work on a "idealised drawing". New questions are now becoming crucial for designers of graphical representations on the screen of a computer. For instance, to what extent are the imperfections of a drawing considered as noise by the users and do not prevent them from having access to the idealized drawing? To what extent is a sequence of small segments accepted as representing a straight line? The computer indeed reveals a phenomenon which is of importance for the pupils in a paper and pencil situation. For the latter the process of eliminating the imperfections of a real drawing is not so spontaneous as for mathematicians because it actually requires mathematical knowledge. Mathematicians know that a circle and a tangent line do not have in common a segment but only a point; they know that the tangent is perpendicular to the radius and they are able to infer this from the drawing even if the angle is not exactly right. But the move of a conceptual nature from the actual drawing to an idealized drawing is not spontaneous for pupils.

An idealized drawing may give rise to several figures depending on the features which are relevant for the problem to be solved. The referent attached to a drawing cannot be inferred only from the drawing but must be given by a text in a discursive way. There are two reasons for this. Firstly, some properties of the drawing may be irrelevant. For instance some relations which are apparent on the drawing may not be part of the figure, for instance the size of the sides a triangle may be not relevant for the problem to be solved, and the position of the drawing in the sheet of paper is usually without any link with the geometrical problem. Only some features of the drawing are relevant for the problem to be solved. And secondly, an important feature of geometrical figures is that they involve elements varying in subsets of the plane. A drawing itself cannot account for the variability of its elements.

The geometrical figure behind the screen

Computers have been used to design programs making apparent the multiplicity of drawings attached to a given geometrical figure. This has been done by several means like programming language or repeat facilities. Improved interfaces now allow direct manipulation while all geometrical properties used to construct the drawing are preserved.

A common feature of these programs is their use of an explicit description of the figures : a drawing produced on the screen is the result of a process performed by the user who makes explicit the definition of the referent. Such programs differ from drawing tools like MacPaint in which the process of construction of the drawing involves only action on the physical screen and does not require a description of the referent, that is of relations between elements. Nevertheless because the design of a geometry software presents some specificities, the referent of a drawing itself cannot be completely identified with the usual referent in Euclidean geometry. In software geometry, the figure is determined by a construction process made of primitives and by the operations which are possible to perform.

A new kind of referent is thus created by such geometry programs, but also new problems may arise because of the novelty of the objects and operations that are possible on these objects. But referents and problems actually differ from one software to another one according to the software design. This question becomes clearer when discussed through the example of Cabri-géomètre.

The Cabri-figure and the Cabri-drawing

Because the drawing made with Cabri-géomètre on the screen has specific features, it will be denoted by the term Cabri-drawing and the object of the theory to which it refers by Cabri-figure. Two kind of primitives are available to make a drawing in Cabri-géomètre:

- primitives of pure drawing which enable the user

   a) to mark a point anywhere on the screen, just at the location shown by means of the cursor (which takes the form of a pen), we call this kind of point a basis point;

b) to draw a straight line while pointing on the screen the position of two points but these points are not created as geometrical objects, we call this straight line a basis line; and
c) to draw a circular line called a basis circle;

- primitives based on geometrical properties which enable the user to draw objects not on a perceptive basis but on a geometrical basis : for example, the user can draw a perpendicular bisector of a segment in selecting the item perpendicular bisector in the menu Construction and in showing with the cursor the given segment.

When a Cabri-drawing is dragged, every constituent of the drawing which does not depend geometrically on the other ones does not move. For example, if a segment AB is drawn and if the user decides to mark its mid-point I only in showing a point correctly located from a visual point of view by means of the primitive basis point, this point I does not move when A or B is dragged (see figs. 1 & 2).

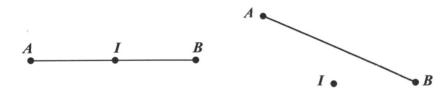

Figure 1: I is a basis point
            visually posed on AB

Figure 2: A is dragged,
            I does not follow AB

In contrast if the mid-point is drawn by using the menu item mid-point its property of being the mid-point of AB is preserved by the drag mode.

The combination of these two kinds of primitives and of the drag mode gives meaning to the notion of figure versus the notion of drawing. A drawing is not resistant to the dragging while a geometrical figure is the set of geometrical properties and the relations preserved by the drag mode. 7 But the theoretical object to which the Cabri-drawing refers presents some features which differ from those attached to the usual theoretical referent. The Cabri-figure is a result of a sequential description process, which introduces an order among the elements of the figure. So in a Cabri-figure some elements are basis elements or free elements on which the other elements are constructed. These free elements can be grasped by the mouse and dragged, they have a degree of freedom equal to 2 whereas a constructed element like the mid-point of a segment is completely dependent and cannot be grasped : its degree of freedom is equal to 0.

Intermediary elements of degree equal to 1 can also be created : these are points on objects, like a point on a circle or on a straight line, they remain on this object when the Cabri-drawing is dragged.

This notion of freedom due to the drag mode and to the necessity of a construction process is not included in the theory of euclidean geometry. It introduces new questions such as what is the trajectory of a point on an object when the Cabri-drawing is dragged?   This is answerable only through the decision of the designers of the software, but which was not completely arbitrary.   The behaviour of the point on an object cannot be too remote from the expectations of the user.   Because a figure in

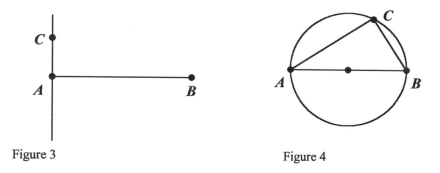

Figure 3   Figure 4

Euclidean geometry is invariant through a similarity, it has been decided to keep the ratio IA/IB constant for a point I on a segment AB.   As a consequence, it would be non relevant to infer properties of a point on an object from observing its trajectory in the drag mode.

The same geometrical figure can give rise to different Cabri-figures. A right triangle ABC right in A can be constructed by choosing A and B as free points and C on the perpendicular line to AB at A (fig. 3) or by choosing B and C as free points and A as point on the circle with diameter BC (fig 4).

Construction aspects and dependence relations play an important role in the way in which the Cabri-figure may be handled. By the way functional aspects of the elements of a figure are emphasized: a perpendicular bisector must be seen as depending on one variable (a segment) or two variables (two points), a perpendicular line must be seen as depending on two variables, a point and a line.

LEARNING GEOMETRY IN A CABRI-ENVIRONMENT:
SOME HYPOTHESES

We believe that the traditional teaching of geometry emphasizes the role of theoretical knowledge and ignores or underestimates the relations between drawing and geometrical theory. Geometry is not presented as knowledge allowing the interpretation of visual phenomena or even the control and prediction of them.

The role of drawings is seen more as an auxiliary illustration of geometrical concepts. Pupils are not taught how to cope with a drawing, they are not taught how to interpret a drawing in geometrical terms, or how to distinguish the spatial properties which are pertinent from a geometrical point of view from those which are only attached to the drawing. In other words they are not taught how to distinguish spatial properties which are necessary from a geometrical point of view from properties which are only contingent in the drawing. This is the reason why there may be some misunderstandings between pupils and their teacher. When the pupils are given a construction task, the teacher perceives it as a task involving the use of geometry, whereas for the pupils it may be a task of drawing. For example, when asked to draw a tangent line to a circle passing through a point P, the pupils often rotate a straight edge around P so that it touches the circle. This is not based on geometry but on perception.

For the same reasons when pupils are asked to prove spatial property, it may be difficult for them to understand why they are not allowed to infer properties directly from the drawing. These behaviours of the pupils have been described by researchers in different countries such as Hillel & Kieran [4], Schoenfield [5], Fishbein [6] and Mariotti [7].

We propose the hypothesis that by designing specific tasks in a software environment like Cabri-géomètre it is possible to promote the learning of geometrical knowledge as a tool for interpreting, explaining, producing and predicting some visual phenomena. And relating to the hypotheses stated at the beginning of this paper, the software environment plays a double role, in that it is the source of tasks and it offers ways of actions and feedback to the pupils solving the task.

NEW PROBLEMS RAISED BY THE COMPUTER ENVIRONMENT

On the one hand, the dynamic treatment of the Cabri-drawings raises new categories of problems in geometry itself, while on the other hand the environment through its specific possibilities allows to give problems to learners which could not be proposed in a paper and pencil environment.

New kinds of problems in geometry

Generecy of a construction. Some constructions depend on the mutual spatial position of elements of the figure which can change under the effect of the drag mode. For example, constructing the tangent lines at a circle from point P outside a circle is not made by the same process as when the point is on the circle. It means that when dragging the point P until sticking it at the circle, the construction process does not provide the tangent line for the end-point of the trajectory of P.

Order of points. When coping only with static drawings, it is difficult to give a meaning to a simple question such as "how to determine by means of a geometrical construction what point among three points on a straight line D is between the two others when these points are moving on D?" It is easy to answer perceptively but the perception provides neither a generic way of solution nor a geometrical process.

Conditional objects. The dynamic feature of a Cabri-drawing may also raise a kind of question which until recently was asked only in a formal setting : how to produce a Cabri-drawing which only exists if the elements defining it belong to a subset of the plane.

The three categories of problems presented here emerge from the fact that the Cabri-drawing is a new kind of representation of geometrical object with specific behaviour produced and controlled by geometrical knowledge:

- it can be moved by direct manipulation (and not by means of a symbolic language) and it keeps its geometrical properties in the movement;
- it is produced by a geometrical algorithm based on geometrical primitives.

This clearly shows that new kinds of representations may give rise to new kinds of problems which call for geometrical knowledge to be solved. Indeed, tasks may be designed for pupils that are specific to the Cabri-environment and which can only be solved by using geometrical knowledge. According to our hypothesis concerning the role of problems solving this kind of task may promote learning for the pupils. What is of interest is to design tasks calling for a use of knowledge which are impossible in traditional environments.

New problems for the learners

With regard to the relations between drawing and geometry, several kinds of tasks can be distinguished:

(i) moving from a verbal description of a geometrical figure to a drawing: this refers in a paper and pencil environment to the classical construction tasks in which pupils have to produce the drawings of geometrical objects given by a verbal description;

(ii) explaining the behaviour of drawings by means of geometry which corresponds to moving from drawing to verbal description and explanation:

   a) interpreting drawings in geometrical terms; this occurs in tasks in which pupils have to prove why a spatial property giving rise to a visual evidence is verified by a drawing;
   b) predicting a visual phenomenon as in problems involving a locus of points;

(iii) reproducing a drawing or transforming a drawing by using geometry.

These three kinds of tasks become different in Cabri-géomètre because of the specificities of a Cabri-drawing. Examples of such tasks are given elsewhere [8], [9], [10], [11] and [12].

Briefly speaking tasks (i) and (iii) should a priori require the use of geometrical knowledge and tasks (ii) seem to motivate the need of proof. All three kinds of tasks involve several aspects of the relations between drawings and geometrical properties

PUPILS' SOLVING BEHAVIOURS WHEN INTERACTING WITH THE SOFTWARE ENVIRONMENT

Although theoretical objects are now reified or embodied in a material environment, the learner will not immediately have access to the meaning intended by the designer of the environment or the teacher. It has often been stated about the Logo turtle that the pupils did not conceptualize the notion of angle simply because they were faced with a task including the construction of angles for instance (Hoyles & Sutherland [13] and Hillel & Kieran [4]. The learners construct a representation of the functioning of the software and of the tasks which are given to them which may differ from the expected representation.

Numerous observations of pupils of grade 8 and 9 working with Cabri-géomètre have given evidence about the pupils' solving processes in this environment, about these use of their primitives and of the drag mode. Briefly, pupils moved from purely visual to geometrical strategies when solving a problem, and the features of the software played an important role in the move from one strategy to the next. It was particularly interesting to observe strategies which mixed visual and geometric steps. At the same time, pupils used the drag mode both as a validity criterion for their constructions, as well as a means of exploration. All observations convinced us that some of the learning was particular to the nature of Cabri, and the software enabled us to observe the process or evolution, of their learning. The evidence and discussion are reported in details elsewhere [8] and [14].

CONCLUSION

Software provides rich environments for modelling or embodying the mathematical objects. The study of the example of Cabri-géomètre in the case of geometry stresses the necessity of analysing a new kind of relation to knowledge which is constructed by the pupils through a software environment. New tasks can be used by the teacher in order to promote learning. But what is their meaning for the pupils? It appears that the visual feedback based on geometry plays an important role in the development of pupils' learning, both by showing the inadequacy of strategies and as giving evidence for some visual phenomena. It also appears that the pupils could use more extensively the conjunction of drag mode and geometrical primitives in order to receive more sophisticated feedback. For this reason, we emphasize the relations between visual and geometrical phenomena through new interface possibilities. In Cabri-like software the scope of visual phenomena is dramatically enlarged and at the same time these phenomena are controlled and produced by theory. They could be used more in the teaching and learning of geometry, and could give a greater meaning to geometrical tasks set for pupils, which up to now are very often viewed by pupils as school tasks without any relation to visual phenomena.

REFERENCES

1. Brousseau, G. (1986) *Fondements et méthodes de la didactique des mathématiques*. Recherches en didactqiue des mathématiques **7** (2), 33-115.

2. Pea, R. (1987) *Cognitive Technologies for Mathematics Education.* In Schoenfeld, A. (ed), Cognitive Science and Mathematical Education, Hillsdale, N J, LEA publishers, 89-122.

3. Parzysy, B. (1988) *Knowing vs Seeing. Problems of the plane representation of space geometry figures.* Education Studies in Mathematics, **19** (1), 79-92.

4. Hillel, J & Kieran, C. (1988) *Schemas used by 12 year olds in solving selected turtle geometry tasks.* Recherches en didactique des mathématiques, **8** (1.2), 61-102.

5. Schonfeld, A. (1986) *Students' beliefs about Geometry and their effects on the students' geometric performance.* Paper presented at the tenth international Conference of the group Psychology of Mathematics Education, London

6. Fishbein, E. (1993) *The theory of figural concepts.* Educational studies in Mathematics, **24** (2) 139-162.

7. Mariotti, A. (1991) *Age variant and invariant elements in the solution of unfolding problems.* Proceedings of PME XV, Furinghetti F. (ed), **II**, Assisi, Italie, 389-396.

8. Laborde, C., & Laborde, M.J. (1993) *Designing tasks for learning geometry in a computer based environment: the case of Cabri-géomètre.* In: Proceedings of the Technology for Mathematics Teaching Conference, Birmingham, U.K. (in press).

9. Capponi, B. (1993) *Modifications des menus dans Cabri-géomètre.* Des symétries comme outils de construction, Petit x, n 33, pp. 37-68

10. Guillerault, M. (1991) *La gestion des menus dans Cabri-géomètre, étude d'une variable didactique.* Memoire de DEA de didactique des disciplines scientifiques, Laboratoire LSD2-IMAG, Université Joseph Fourier, Grenoble

11. Bergue, D. (1992) *Une utilisation du logiciel "Géomètre" en 5ème.* Petit x, IREM de Grenoble, **29**, 5-13

12. Boury, V. (1993) *La distinction entre figure et dessin en géomètrie : etude d'une "boîte noire" sous Cabri-géomètre.* Rapport de stage du DEA de sciences cognitives, Equipe DidaTech, LD2 IMAG, Université Joseph Fourier, Grenoble

13. Hoyles, C. & Sutherland R. (1990) *Pupil collaboration and teacher intervention in the LOGO environment.* Journal für Didaktik der Mathematik, **11** (4), 323-43

14. Berthelot, R & Salin, M.H. (1992) L'enseignement de l'espace et de la géomètrie dans la scolarité obligatoire, Thèse de l'Université Bordeaux 1.

Colette Laborde graduated in mathematics from the Ecole Normale Supérieure, obtaining a "Doctorat es Sciences" in mathematics education at the Scientific University of Grenoble, specialising in language problems in mathematics teaching and learning. She is a full professor at the University Teaching Institute and a member of the mathematics education research group at the Institute for Applied Mathematics and Computer Sciences (IMAG). Colette Laborde is head of the doctoral programme of mathematics and science at the University of Grenoble and currently a full researcher of the CRNS.

Jean-Marie Laborde graduated in mathematics from the Ecole Normale Supérieure in Paris and started his research work at the Institute for Applied Mathematics and Computer Sciences (IMAG), specialising in the use of geometric methods for the study of different classes of graphs, especially hypercubes, and also automatic theorem proving. With others he started the Cabri-géomètre project in 1981, initially as an environment for graph-theory. In 1987 a number of students and young researchers joined the project to start Cabri-géomètre, which a year later was nominated as Educational Software of the Year by the Apple Company. In 1982, Jean-Marie Laborde founded the Laboratory for Discrete Mathematics and Research at Grenoble University. He has been guest professor at several universities and is currently Research Director of the CRNS and Head of the Cabri-géomètre Project at IMAG.

10

Moving schema

Ichiro Kobayashi

Kawaijuku Educational Institution
Tokyo, Japan

ABSTRACT

Just as a Venn diagram can represent both a loop and a mathematical set, semi-concrete schema function as a bridge between the concrete and the abstract and help students to understand mathematical concepts. Using computers, schema can be moved in real time by the teacher to clarify the object in question, to make comparisons of the object before and after transformation, and to focus on invariants. In motion, schema take on greater strength as signifiers, and become harder to understand intuitively. Our task now is to learn how students can come to visualize moving schema for themselves.

Keywords: graphics, teaching methods, thinking

INTRODUCTION

This paper provides me with an opportunity to discuss the importance of moving schema in mathematics education and the role of computers in facilitating this concept. It is an area that has yet to receive the attention it deserves and one that I hope will be debated by more people in the future.

I use the word 'schema' here to denote a simple figure used to illustrate a mathematical concept. This figure is somewhere between a concrete object and an abstract object; it might reasonably be called a semi-concrete object. It is similar to what in psychology is referred to as a 'mental model', but for this discussion 'schema' denotes both a mental image and a real figure.

As computers develop, it is becoming easier and easier to create, move and display such figures, and we can expect moving schema to continue growing in significance. Moving schema have had a place in the imagination of mathematicians for some time, but have yet to acquire a prominent role in education, perhaps because they are not easily taught to students. One thing is certain, however, and that is that schema are not primarily a private means of solving problems - they are communicative devices that can speak to many people.

THE ROLE OF MOVEMENT

Qualitative understanding

As Piaget said, 'Qualitative understanding precedes quantitative understanding', and moving figures created with computer graphics are a highly effective tool for leading students to a qualitative understanding of mathematical concepts. Students, for example, have difficulty when they are asked to obtain the eigen vector of two-dimensional linear transformation using defined expression and calculation. On the other hand, when they are shown with computer graphics the transformation of radiating lines passing through the origin, or the transformation of a number of points on a plane, students can sense the existence of the eigen vector even though they do not know its precise value. Offering students a qualitative understanding first leads them to an understanding of defined expression and gives them a motive to calculate answers for themselves. This is especially true in the case of a special matrix, such as $\begin{pmatrix} 3 & 2 \\ 2 & 3 \end{pmatrix}$, where students can only find the eigen vector $\begin{pmatrix} 1 \\ 1 \end{pmatrix}$ after visualizing the transformation. This is shown in Figures A1 through B2.

This is a technique I have used successfully in many of my own classes on the eigen vector. Six years ago, however, my first attempt ended in failure. It was a summer class on basic linear transformation, and it failed because I mistakenly believed that every student would automatically perceive the relationship between the crossed stripes being transformed on the screen and the linear transformation itself. Not everyone did: some students saw in the image nothing more than figurative movement; they did not see the image as the signifier of a

$$\begin{pmatrix} 3 & 2 \\ 2 & 3 \end{pmatrix}$$

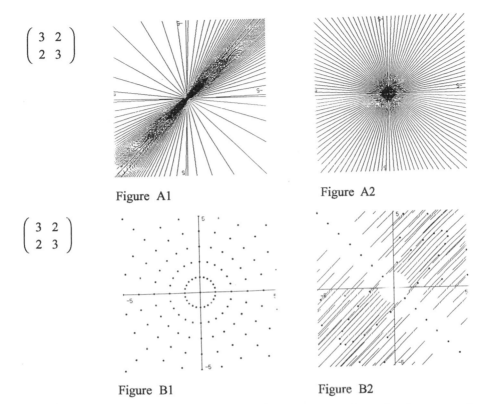

Figure A1

Figure A2

$$\begin{pmatrix} 3 & 2 \\ 2 & 3 \end{pmatrix}$$

Figure B1

Figure B2

mathematical concept. To do this students must analyze the image and teachers have to help them in this regard. My solution after that experience was to put an indicator arrow in the picture that I could move around with the mouse. I find it beneficial to first have students consider the image of $\begin{pmatrix} 1 \\ 0 \end{pmatrix}$ and then to examine the image of $\begin{pmatrix} 0 \\ 1 \end{pmatrix}$. In this way, they come to understand the image as a whole and can get used to viewing that image as the signifier of a concept. (See Figures C1 and C2).

$$\begin{pmatrix} 3 & 2 \\ 1 & 4 \end{pmatrix}$$

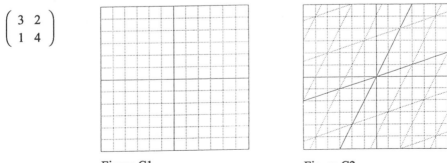

Figure C1

Figure C2

Inner product is another example. This is a difficult concept and students frequently have problems in understanding how to calculate it. In this case, I offer them near-concrete examples of rain and an umbrella, where both are seen as vectors because they have direction and size, and where the volume of rain striking the umbrella can be expressed using inner product. Here, students are shown that they do not need to concern themselves with concrete values. They see that changing the direction of each element changes the volume of rain that strikes the umbrella, and can accept the necessity of considering inner product as volume. Once students reach this stage, they are shown that inner product can be shown simply in terms of the area of a rectangle and come to a more complete understanding of the concept.

Semi-concrete objects

The schema of inner product I have just mentioned can be seen as halfway between a concrete object and an abstract figure. This can also be shown in another example, that of an explanation of $\sin\theta$. Students are asked to imagine a crane ship, floating on the water. They are told that the angle between the crane and the water's surface is θ, and that the distance separating the tip of the crane and the water's surface is $\sin\theta$. They are then asked to notice the changes in the crane's height as the angle is altered and, following the movement of the crane, they can estimate the value of $\sin\theta$ when the angle is between θ and 180°. Of course they see only a rough value, but their estimations lead them to an understanding of the meaning of $\sin\theta$ (see Figures D1 through D5).

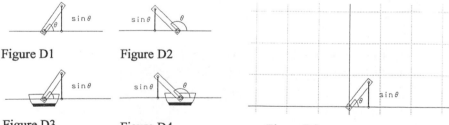

Figure D1   Figure D2

Figure D3   Figure D4   Figure D5

Following this, co-ordinates are added to the figure in the plane whose origin is the ship, and students are brought closer to an understanding of the abstract concept of $\sin\theta$. They are then taught the definition of a unit circle, and are shown an image of a unit circle in a black box. An angle is entered and the value of $\sin\theta$ is given by the unit circle. Students can see the output image of the value of $\sin\theta$ as a value of function. Certainly, we can regard the unit circle itself as the schema for trigonometric function, but by moving the figure, teachers can focus more effectively on the process as a function.

We can also show the composition of trigonometric function through the revolution of folded lines.

$$a \sin x + b \sin (x + \theta)$$

When $\theta = 90°$, we can describe the composition of $a \sin x$ and $b \cos x$, and when $a = \cos \alpha$, $b = \sin \alpha$, we can show the additive theorem (Figures E and F). Moving schema are invaluable in areas like these: they allow teachers

$$y = 2 \sin x + \sin(x + 90°)$$
$$= 2 \sin x + \cos x$$

$$\sin(\alpha + \beta) = \sin \alpha \cos \beta + \cos \alpha \sin \beta$$

Figure E   Figure F

to demonstrate difficult concepts with simple figures and movement. The following example (Figure G) uses the form of a snail shell to teach exponential function. The spiral of the shell is presented as a particularly elegant form of this function and is shown at first in natural section before being reduced to a simplified spiral. It is then used to explain logarithmic and exponential law.

The three cases described here may be successful examples of the use of moving schema: their abstract renderings of concrete objects have certainly proved consistently popular with students.

$$y = a^x, \quad a = 3$$

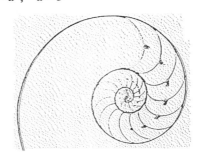

Figure G

Movement and preservation

Movement is useful in teaching laws and in focusing on elements or factors which remain constant. If we return to the previous example of inner product, we can consider, say, the inner product of a fixed vector and of a variable vector (see Figures H1, H2 and H3 opposite), and we can establish that only when $\vec{x}$ moves perpendicularly to $\vec{a}$ does the inner product remain constant. We can then see the equation of the straight line. In other words, when $\vec{a} = \begin{pmatrix} a \\ b \end{pmatrix}$, $\vec{x} = \begin{pmatrix} x \\ y \end{pmatrix}$, the equation of the straight line passing through the point (x_0, y_0) is

$$\begin{pmatrix} a \\ b \end{pmatrix} \bullet \begin{pmatrix} x \\ y \end{pmatrix} = \begin{pmatrix} a \\ b \end{pmatrix} \bullet \begin{pmatrix} x_0 \\ y_0 \end{pmatrix}$$

$$\Rightarrow ax + by = ax_0 + by_0$$

The whole and the parts

Using two arrows and the area of a rectangle, teachers can show two vectors and the inner product of $\vec{a}$ and $\vec{x}$. In this way, when students are shown how to change the inner product by moving the arrows, they can readily appreciate that the value of the inner product affects each point of the plane. This is made clear by the shaded area of the rectangle, and is a clear illustration of how examining the part (the straight line) in relation to the whole leads to improved understanding overall.

In the same way, we can consider the example of linear transformation with a determinant of 0.

The nature of this linear transformation is clearly revealed when the straight line $x + y = c$ is shifted to the point (c, c) and the plane as a whole transforms into the straight line $y = x$. If students first look at the whole image and follow the transformation of each point on the plane, understanding how a part of the image, again a straight line, is transferred to a point becomes much simpler. Computer graphics aid understanding still further because they can also show the whole image in motion.

These principles also apply in the area of functions. Students can better understand the graph of, say, $y = e^x \sin x$ by first considering the whole graph of $y = a^x \sin x$. Here, when $a = 1$, $y = \sin x$, as in the original form, and they can see that with successive changes to the value of a, there is a certain moment during the graph's movement when $y = e^{-x} \sin x$. After considering $y = xa^x$, they can accept that $y = xe^{-x}$ can be obtained from $y = x$. Viewing graphs of functions in this way, that is, as shifting curves on planes, may be an important means of improving understanding (see Figures I1, I2 and J).

$$\begin{pmatrix} 4 & 0 \\ 1 & 0 \end{pmatrix} \bullet \begin{pmatrix} 1 & 8 \\ 6 & 6 \end{pmatrix} = (13, 8)$$

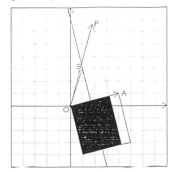

Figure H1

$$\begin{pmatrix} 4 & 0 \\ 1 & 0 \end{pmatrix} \bullet \begin{pmatrix} 0 & 2 \\ 6 & 2 \end{pmatrix} = \begin{pmatrix} 7, 0 \end{pmatrix}$$

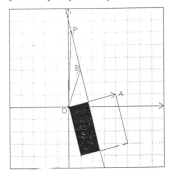

Figure H2

$$\begin{pmatrix} 4 & 0 \\ 1 & 0 \end{pmatrix} \bullet \begin{pmatrix} -0 & 6 \\ 6 & 0 \end{pmatrix} = (3, 6)$$

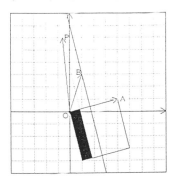

Figure H3

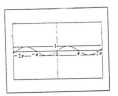

$$y = \sin x$$

Figure I1

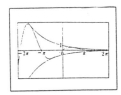

$$y = a^{-x} \sin x$$

Figure I2

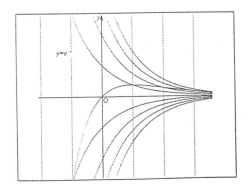

Figure J

Experience has shown that using examples like these in short teaching periods is ineffective. Perhaps students need time to establish the images in their own minds; certainly using these images does not mean that students will suddenly grasp the concepts involved. Symbolizing concepts brings creates its own difficulties, but immediate understanding is not necessarily a good thing in this area: students need time.

The point at issue
The ability to analyze is the key to seeing graphs as connections which carry meaning and as signifiers of concepts Students do not perceive these images in the same way as teachers, and need to study a large number of them before they can acquire the necessary analytical skill to understand them correctly. Without it, they will continue to misunderstand the movements within figures and fail to distinguish certain movements from others. For the moment, I have little idea of how student analytical proficiency can best be developed.

CONCLUSION

Moving schema created with computers can enhance students' ability to form such schema in their own minds, and play an important part in understanding concepts. The most pressing task for the present is to order and arrange these figures in terms of movement so that students can understand them more easily. Moving schema, using the dynamic graphical capabilities of information technology are enabling mathematics students to tackle higher order conceptual thinking.

Ichiro Kobayashi has been a lecturer in mathematics at Japan's Kawaijuku Educational Institution since 1987. His publications include *Calculus for High School Students* (Sanseido) and *Linear Transformation of Second Degrees for High School Students* (Obunsha).

11

Developing analytical competency through informatics teaching

Walter Oberschelp

Lehrstuhl für Angewandte Mathematik (RWTH),
Aachen, Germany

ABSTRACT

A short summary is given of informatics teaching in Germany. As goals for informatics education, some competences are listed which can only be achieved through informatics. This leads to a justification for obligatory informatics courses in secondary instruction in Germany. Such teaching must be supported by tools, which have certain fundamental requirements: zooming in time and space, algorithmic transparency and a simplicity of interfaces. Graphical facilities using the discrete pixel plane become necessary components of software, which is available now or shortly. Onc major concern is to prevent the degeneration of algorithmic capabilities as a possible consequence of software that is too easy to use.

Keywords: tools, algorithms, graphics, inter-active, instruction

INTRODUCTION

As an introduction, it is important to provide a short resumé of the experiences of teaching informatics in German high schools. This is followed by comments and personal opinions on recommendations for the curriculum [1]. The author has been one of the members in the committee preparing these recommendations, based on the so called Dagstuhl-Recommendations [2].

It is my opinion that the experiences in Germany with basic courses on computer literacy for all pupils, called ITG (Informationstechnische Grundbildung) have not been convincing. There was neither a general consensus on the curriculum nor did we get any realization line for a broadly based teaching system for ITG. The attempt to make mathematics teachers responsible for this kind of instruction met with resistance from all groups, and there was no general agreement on fitting an ITG-block into the curriculum (for conflicting positions see [4] and [5]). But, the impossibility of achieving ITG is fortunately connected to the fact that ITG seems not to be really necessary: the mere skills of handling a computer must not be the object of school instruction at all, any more than the skills in car driving, typewriting or handling another complicated technical tool. The trend that makes computer interfaces more easily understandable, as a consequence of the marketing needs of industry, combines with the experience that our children do not fear the computer. Other aspects of ITG, in particular the reflection on social implications, can be covered by conventional educational components, since the substance is too thin for special courses with only a small and elementary background.

On the other hand, my proposition is that there should be compulsory courses in informatics at least in each curriculum leading to college maturity. This position is not generally accepted (see [6] and [7]). The development of high school informatics in Germany over recent years shows a somewhat chaotic situation as a result of different cultural policies in our federal system. In general, a big boom has occurred, with many teachers - mostly mathematicians - having undergone additional and voluntary training in informatics. Thus, in many schools a high quality instruction in computer programming and programming principles is available, but normally on a free choice basis. Virtually nowhere is informatics compulsory. A good description of the situation and some references are given in [1]. More details can be taken from [8] and [9], while recommendations for the teachers' education are given in [10].

There appears to be growing competition with other fields, mainly with mathematics, not only about curriculum proportions, but also about the content of instruction. Mathematics seems to become more and more aware of the fact that lost fields have to be regained - in particular the whole world of algorithms. There are also new challenges. The advent of formula manipulation systems into mathematics instruction will cause a revolution, as the invasion of high-level software tools into informatics seems to make many "classical" applications obsolete.

GENERAL PRINCIPLES OF INFORMATICS INSTRUCTION

It would seem that informatics instruction has to change towards the future with more maturity. At the present state of technology it is possible to make recommendations which are rather independent from the actual availability of those tools. It is important to keep in mind and to insist on some educational principles which might be unpopular and uncomfortable from the aspect of those advanced tools.

The justification of this opinion is based on the fact that there are special competences for our information based society which cannot be developed in other fields of instruction. In general, in order to master the new opportunities of information processing, it is necessary to know how to use these opportunities and to understand the fundamental processes which are going on in complicated systems, on both the elementary and more complex levels. Equipped with those competences, pupils will be able to use, analyze and evaluate information and to create their own solutions for special problems later, thus preparing for autonomous positions in our complex world. The competences, which need to be developed, are listed as follows.

- recognize and analyze problems, which are solvable appropriately by a computer, (observation competence), in particular
     a) recognize language (natural or artificial text) - as the fundamental vehicle of communication and of interactive structured problem solving
     b) use visualizations and pictures and their inherent patterns as an aid for perceiving structures and for recognizing correlations between facts;
- deliver precise and appropriate (re)actions on a syntactically and semantically sound basis in interaction either with linguistic or graphical representations of a software tool or with partners of a working group (serving competence);
- apply those abilities to various fields and design sequences of actions either using a software tool in order to solve a problem (programming competence) or working autonomously or as part of a group (combinational competence);
- recognize the role of basic low-level-activities in simple and complicated settings and practice those activities with or without the support of a software environment (algorithmic competence);
- find quantitative estimates for complex processes, distinguish essential parts and judge the role of communication and interfaces between those components (evaluation competence);

- combine, with the help of appropriate tools, components, which have been, or can be, fully evaluated in a systematic and incremental way to a completed structure (assembling competence);
- verify the functionality of a construction by efficient checking procedures (test competence);
- improve solutions by appropriately adapting the requirements to each other (optimizing competence);
- distribute tasks and find parallel solutions bearing in mind communication needs between processes (parallel design competence);
- estimate human conditions and social contexts in the classroom and in a creative working group and develop responsibility (personal competence).

These competencies should form the basis of informatics instruction.

SOFTWARE TOOLS - REQUIREMENTS AND PROPERTIES

These general goals of informatics instruction affect both the tools and the subjects of our instruction. The software and hardware tools of instruction must be selected and developed according to their value as educational devices. The usual evaluation of software favours technological perfection for direct and fast solutions. This has also been the case in German institutions, which evaluate software for instruction purposes, as I know from personal communication with K. Keidel (Augsburg) of the Bavarian test institute. Of course, we want to follow technological progress; but usually advanced commercial software is not suitable for instruction, since transparency and motivation are lacking. We should encourage the use of those tools, which show intermediate steps, support imagination and analytical endeavours and which are equipped with flexible interfaces. In contrast to the fixed special purpose interface for the private and commercial user, students should be able to communicate with the system or with their teams through a flexible or at most semi-automatic interface. We know from experience that working with such interfaces challenges improvisation and needs robustness and intensive dedication. Therefore, in order to prevent an isolation of students, such that the computer is their only communication partner, the teacher often schedules an instruction unit that co-ordinates the work in advance by informal agreements, as support through the system itself cannot and should not be relied upon.

The big achievement of the computer is its ability to transform any action of a user directly into a reaction. Thus users have direct evidence that their actions cause an effect. The most direct immediate contact with the medium is the interactive screen in handling menus, texts and graphics. Working with the editor in an interactive programming environment, for example, a text system editor, requires lexical and syntactic correctness. Thus the editor is a teacher by itself. The supervision function of the system insists on absolute exactness, it gives hints to localize errors and forces a search for the error with more or less support. The position of the cursor supports precision in sequential thinking, it makes conscious the fundamental state-space principle, that is the system is at any time in an actual state (within a partially hidden context) and the cursor reflects the activity of the user, who works with comfortable input devices.

It is obvious that students also have to know the semantic meaning of an action. They have to master a minimal repertoire of meanings for procedures on data structures, such as sort, search, retrieve, select, cancel, compare, count, map, merge, collect, compile, concatenate, and transitive closure. This and the knowledge of the fundamental logical and arithmetic functions are an indispensable subject of instruction and must be covered by any curriculum.

ALGORITHMIC SKILL AND TIME AND MOTION CONSUMPTION

In this context it is important to evaluate critically formula manipulation systems. Experiences in Germany and Austria [11] have shown, that a considerable acceleration in mastering the mathematics curriculum can be achieved and that in particular the teacher, who is designing classroom tests, can take advantage of such systems. But on the other hand the need to formulate exactly the desired goal sets a clear limit for applications without precise ideas, and we have formulated (cf. [12]) the danger of degeneration for the permanent user of a formula manipulation system. Thus, in the same way that the lack of use of muscles can lead to physical degeneration, the use of perfect automatic tools may lead to mental atrophy and to an algorithmic disabling. I will return to this point later in the paper. There is no clear position among teaching authorities in Germany and as yet a lack of consciousness in the public discussion. But this is not unexpected; after all the discussion about value and limitations of the pocket calculator in instruction has not yet reached a conclusion. While mathematical rigour and logical precision is the hard core of mathematic instruction and competes with numerical calculations, the ability to perform algorithmic manipulations and thus to experience the resulting cause-effect-relation is an indispensable informatics quality, which cannot be substituted by even perfect software tools.

It is well known that the attraction of using such an advanced tool without any exertion, like a recreational toy, is one of the main motivations for our youth to be occupied with the computer. Students are not to be blamed, therefore, but they have to make some progress and move from playing with the tool to challenging it, in order to look behind the apparent view of the action. Thus, for example, the question "what is going on algorithmically if a little man is moving over the screen, if it is suddenly disappearing or created ex nihilo?" should be discussed as well as the question how to manage the creation of straight lines, circles and nice curves on the screen.

In my opinion is a crucial point; it is important to develop strict ideas about the conditions for "implementing" motion on the screen. It should become clear to the student that within 50 msec - this is the critical time to get the impression of a continuous motion - a 20 MIPS-computer can perform one million instructions. It is easy to build a simple figure on the screen, it is possible to change the direction of a drawn line within that time, if we know what to do in advance or if the system has to react to a mouse click. But all this is impossible if the action is triggered by the result of a combinatorial calculation like the inspection of all possible orders of ten objects or on a logical condition, where twenty binary variables are involved. Most students in Germany have no experience in discrete optimization and thus are too innocent to take into account the influence of the combinatorial explosion on computer performance. But I feel that for a software tool the transparency of the performance time is a most important quality. All our tools should be equipped with a device to achieve time-zooming in order to expose discrete motions, to use a step by step mode and to control a speed-down and, if possible, a speed-up by continous parameter changes. In many cases this is a rather harmless extra feature of the software, but it is indispensable for learning purposes.

There are important examples to illustrate the time problem in informatics. One classical, well elaborated project [13] is the one-way (trapdoor) calculation problem, that is the totally different time requirements to multiply two big numbers and to factorize a given number. We use this facility for achieving data security. It is similarly possible to study the tricks of a logarithmic speed up by divide-and-conquer techniques in order to improve sorting, searching and number crunching.

CONSTRUCTIVE CREATIVITY FROM ALGORITHMIC CAPABILITY AND PERCEPTUAL LEARNING

In general, the amount of material which is important for informatics instruction is tremendous. But the most important issue is the way in which the materials enable learning. The learning process to create, design and implement a complex solution by the structuring, incrementally supplementing, prototyping, implementation, testing and documentation can happen in different environments. The students can work in a cooperative situation with a real world team, or in a network of processors or with an interactive software tool. In the latter case the reliability of the components is within the scope of only one's own responsibility; in other cases the student has to estimate the reliability of the components or of the companions. Both situations are highly educative and should be practiced in various situations.

What is the basic ability a student needs in order to achieve all this? Students need to develop the ability to use informatics as an appropriate access to create artificial worlds and to simulate real-world-actions on the computer. But teaching informatics cannot only mean to teach the use of those complex tools and to work in self-created maybe highly sophisticated artificial worlds. It is also important to guard against the danger of loosing reality and becoming a somewhat autistic hacker without contact with the real life.

On the other hand, it is also important to fight against the degeneration and shrinking of vital and necessary abilities by insisting on training the natural gifts not only of our body but also of our mind. In the same way, as we preserve physical fitness by training and by forcing the student to produce sweat, we should have a concern to produce "mental sweat" by occasionally and on purpose stopping the computer and running things on our own brain. This might be the most unpopular part of our education business, which is as hard to motivate as making a man run with his own feet when he usually moves by car or plane. In the same way as we have to preserve literacy by not only relying on pictures or to maintain numeracy by sometimes avoiding the calculator, we have to preserve logicacy and algorithmacy by continuously training simple mental activities without artificial tools. The development of realistic training programs and a good coordination to the other parts of informatics instruction is one of the most important curriculum concerns at all. I have the impression that these needs are not yet recognized in our school system.

To return to the discussion of materials and methods, one important chance for informatics instruction is the use and development of simulations; simulating a system in particular with stochastics-support is the most direct application of the immediate cause-effect-visibility paradigm, and simulation is a fundamental technique for system development. Thus, for instance, modelling the ecological balance of a lake by means of a predator-prey simulation can make transparent the behaviour of linear systems, of non linear feed-back-controlled systems, under the logistic equation, and even of chaotic systems [13]. Here, the crucial topics are linear systems in their temporal development.

I emphasize again that the importance of real time can hardly be over-estimated in informatics instruction. The puristic concepts of top-down or bottom-up techniques turn out to be no longer necessary under modern software tools, since incremental working very often helps to avoid several passes through the system, which are usually necessary in systematic system design. But in this context my former remarks concerning degeneration through luxurious tools also apply. The minimal requirement for using a database-tool, for example, is to make the structure of such a system visible, to show access-paths, reveal with painstaking exactness the reasons for concurrency conflicts. It is important to visualize the actions, which happen during the execution of a select-from-where- statement such as merging, selection, tests, decision trees. The same is true for learning to understand the actions which are going on in computer hardware at the binary (logical) level [14]. There exist already excellent visualization tools which illustrate these processes and which can be applied, when the basic low-level facts have been mastered. I know of several such solutions, which have been developed by teachers with good results in classroom instruction. There is much discussion on this visualization point. The system Cabri-géomètre [15] and [13], is a useful contribution here, which is used in several places in Germany in order to make geometric notions more dynamic and thus more evident. The already historic turtle-geometry technique also reminds us that motion in a graphical setting is one of the basic fundamentals of perceptional learning.

THE PARADIGMS OF DISCRETENESS AND OF PARALLELISM

Why do I concentrate the discussion to these temporal aspects? It should be clear that for most people the main source for cognition is visual perception on the screen. I feel that students sometimes are deceived by its misleading high-resolution. It is important to emphasize that the screen is a discrete and pixel-oriented device. In many cases this should be demonstrated by making the raster visible through zooming. Informatics with its discrete philosophy has to fill the gap which arises from the usual neglect of discrete facts in mathematics instruction. Unfortunately, discrete pixel geometry is more complicated and not as nice as Euclidean Geometry compared with reality. There are also many algorithmic needs in computational geometry under Euclidean assumptions - thus computer algorithms revitalize a lot of classical long-forgotten mathematics. Being also a mathematics teacher, I want to formulate here my belief, which I know annoys many computer scientists, that informatics has the closest and most natural connections to mathematics and that we saw off our own position-branch if we cut off the connections to mathematics.

Some important spatial notions are better understood in the discrete manner. Most important seem to be the concepts of locality versus globality: the use of templates, that is, the processing of a pixel P by a function which depends on the neighbourhood of P, is the most effective algorithmic technique of graphics and of image processing, which can eventually also yield global informations about a picture P, like size, connectedness and genus [16]. The most attractive feature of those local operators in graphics is their use in the parallel mode - the inertia of the eye and the speed of the computer makes it simulate actions as parallel as if they are done by one processor sequentially. Thus thinking and designing in parallel is supported by visual perception on the two-dimensional screen, and this ability also works as the basis for pattern recognition with important applications in medical diagnostics. I am sure that in a very short time the equipment for informatics courses will contain the facilities of a graphic station.

Thus the discussion finally moves to parallel competence. School instruction has the chance to create and to further develop, for example, the natural ability to comprehend the working of a pipeline or more general of a systolic array. It is not easy to design and to verify such a fundamental device of technique. From a theoretical point of view it seems attractive also to deal with fundamental protocols of parallelisms which handle, for example, synchronization, distributed access, mutual exclusion or deadlocks. All these phenomena appear in our society from planning of production to the simulation of social processes. It seems to me, however, that we must restrict ourselves usually to study so called SIMD-processes (single instruction on multiple data) like mesh-connected or hypercube-calculations or multi-connective nets like perfect-shuffle, Butterfly-, Benes- or Clos- structures [14]. In my opinion there is no other field of education than informatics, where all these fundamental materials can be learned.

HOW CAN WE ACHIEVE ALL THIS?

Finally it is necessary for us to consider the position of informatics in the general process of education (in German we use the non-translatable word *Bildung*). I have attempted to identify genuine and irreplaceable qualities of this instruction, as well as examples where informatics touches other fields. On the border with the linguistic sciences we have the concept of syntactical structure and the need to formulate ideas with appropriate semantical notions. We touch the historic sciences in analyzing and creating temporal processes considering contextual state conditions and taking into account necessary, expected and possible actions. The frontier with mathematics is the competition on algorithms and in modelling, with physics the analysis and handling of strict quantitative laws, with biology the technique to represent knowledge and to work with evolutionary concepts. The future of informatics teaching at school will require a recognition of these specific features of our field and to transfer them into the education process by designing a reasonable curriculum.

The way to achieve this competence in a one year course can hardly be sketched here. Starting from a medium position, that is neither top nor bottom, which is close to the actual situation of the pupil and using commercial software cautiously, some examples or paradigmatic small theories can be pursued incrementally in different directions. This can lead to a partly tree-like, partly meshed experience with concrete and important material. The examples would have to be selected with great care so that a representative block of know-how, which is valuable for our educational purposes, is accomplished. Some detailed descriptions are in our paper [3], but more of the advanced tools which we have mentioned must also be included.

The choice of material should be decided through national or even international discussions. The day that interested teachers create their own material for their purposes should be over, while still allowing some room for the creativity of the teacher. I think that such a common basis of knowledge, selected by admittedly somewhat arbitrary arrangements, might nevertheless be an appropriate basis for the development of informatics abilities and for the acceptance of informatics in the general curriculum.

REFERENCES

1. Fäkulttentag Informatik (1993) *Empfehlungen zum Schulfach Informatik (Sek. II) und zur Ausbildung von Informatik-Lehrkrften.* (Informal paper).

2. Claus, V. et al (1992) *Dagstuhler Empfehlungen zur Aufnahme des Faches Informatik in den Pflichtbereich der Sekundarstufe II, Dagstuhl.* (Informal paper).

3. Oberschelp, W. (1987) *Algorithmen und Computer im Unterricht (Projekte).* Fernkurs an der Fernuniversität Hagen (Fachbereich Mathematik und Informatik).

4. Baumann, R. (1990) *Didaktik der Informatik.* Klett, Stuttgart

5. Penon, J., Sack, L. & Witten, H. (1992) *Informationstechnik und Allgemeinbildung.* LOG IN **12**, 22-28

6. MNU (1993) *Positionen der MNU zum Unterricht in Mathematik,* in den Naturwissenschaften und in Informatik, MNU 46 , Heft **8**

7. DMV (1983) *Stellungnahme zum Informatikunterricht an Gymnasien.* Dt. Math. Vereinigung, DMV, MNU **36**, 304 ff

8. Baumann, R. & Koerber, B. (1991) *Informatik in der Schule der 90er Jahre.* LOG IN **11**, 31-35

9. Peschke, R. (1990) *Grundideen der Informatikunterrichtes.* LOG IN **10**, 25-33

10. (1987) *Gesellschaft für Informatik GI,* Empfehlungen zur Lehrerbildung im Bereich der Informatik. Printed in LOG IN **7**, Heft 5/6.

11. Weigand, H.G., & Weth, T. (1991) *Das Lösen von Abituraufgaben mit Hilfe von Derive.* MNU **44**, 177 ff

12. Oberschelp, W. (1988) *Rechenverfahren und Formelalgorithmen als Unterrichtsgegenstände,* in: Arbeiten aus dem IDM der Universität Bielefeld, Occasional Paper **116,** Koll, W. & Steiner, H.G., ed. Winkelmann, B.

13. DIFF (1988) *Computer im Mathematikunterricht,* Dt. Institut für Fernstudien DIFF Tübingen, Beltz, Weinheim Steiner-Colloquium, Winkelmann, B (ed).

14. Oberschelp, W. & Vossen, G. (1994) *Rechneraufbau und Rechnerstrukturen,* Edition 6, München (Oldenbourg)

15. Schumann, H. (1990) *Neue Möglichkeiten des Geometrielernens durch interaktives Konstruieren in der Planimetrie,* MNU **43**, 230 ff

16. Oberschelp, W. (with Dohmen, M.) (1991) *Mathematische Methoden für Bildverarbeitung und Computergraphik, Schriften zur Informatik und angewandten Mathematik,* RWTH Aachen, Nr. 149

Walter Oberschelp was born in Herford, Germany, studying at Göttingen, Tübingen and Münster, gaining his PhD in Mathematical Logic in 1958. After holding professorships at Münster, Hannover and Illinois, he was appointed to his present post as Professor of Applied Mathematics and Informatics in Aachen where he has developed the Informatics Department and engaged in scientific work. He has delivered courses on informatics, discrete structures and algorithms. Walter is also an advisor for many student teachers in mathematics and informatics, giving courses in high school instruction for both students and teachers.

12

Integration of informatics into education

Tom van Weert

School of Informatics
University of Nijmegen, The Netherlands

ABSTRACT

Pushed by technology and pulled by society, Informatics and its applications are penetrating our lives at great speed, with considerable effect on work, education and leisure. Educational systems appear to lag behind technological developments. Demands from society, however, will force education to integrate information technology, and also Informatics as a science. Students will study in a multi-disciplinary team environment supported by integrated information technology. And a new Informatics Literacy, based on Discipline Integrated Informatics, will emerge.

Keywords: informatics as a study topic, integration, interdisciplinary, learner centred learning, literacy.

INTRODUCTION

Informatics or Computer Science is a branch of knowledge or study, especially concerned with establishing and systematising facts, principles, and methods, as by experiments and hypotheses. Its field of study, the design and realisation of programmable systems, is depicted in figure 1.

As Pure Informatics it is pursued simply oriented towards itself, without reference to applications. As Application Oriented Informatics it is pursued as a science oriented towards the design of applications. All of

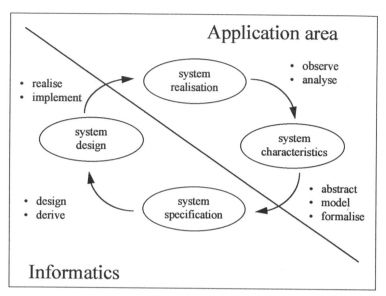

Figure 1: The design and realisation of programmable systems

such applications in society constitute a technology: Informatics Technology.

Sometimes the bond between informatics and another science or technology is so close that the two melt together into an "Informational Discipline" or an "Information Technology". The term Information Technology is often used to denote all the combinations of Informatics Technology and other, closely related technologies.

THE INTEGRATION OF INFORMATICS INTO HIGHER EDUCATION

In the curricula of higher education informatics is found in several forms: Pure Informatics, Application Oriented Informatics, and as part of Informational Disciplines into which informatics is integrated [1]. Learning to use applications of information technology is an integral part of study activities and is in general not considered to be part of an informatics study.

Pure informatics

Pure Informatics, the study in which informatics is studied purely for its own sake, encompasses such topics as automata theory, compiler construction, operating systems and complexity.

Application Oriented Informatics

Informatics may also be studied with the aim of eventually building application systems, which are currently found in four broad, economically important areas. These are:

• Business Oriented Informatics deals with the development of information systems for applications in support of secondary business processes or, in the service industry, in support of primary business processes. It concerns itself with information needs of organisations.
• Technical Process Oriented Informatics deals with the development of technical systems for applications in the primary industrial processes, such as process control, telecommunications and computer aided design.
• Language Oriented Informatics deals with the development of language systems, language interfaces and with speech production and recognition.
• Cognitive Informatics deals with the development of knowledge systems and ergonomic human-computer interfaces.

These four types of informatics are each characterised by a special-purpose knowledge base and special purpose methods, techniques and technology. There is, however, a core of Application Oriented Informatics which is common to all four application areas; elements of this core are modelling, programming, design of data structures, software engineering and human-computer interfacing.

Informatics integrated into "Informational Disciplines"

Some scientific disciplines are strongly bound to informatics. This bond is forged by the use of principles, models, methods and techniques from informatics. The role of informatics in these disciplines is not just limited to the use of applications of informatics, but extends into the methodology of the discipline. The scientific questions addressed are within the domain of study of the discipline concerned, but their solution depends on scientific elements of informatics. Generally speaking, an informatics specification of possible answers to questions is developed, which then may run as a program on a computer. In other words, in an informational discipline, problems are analysed and solutions developed as dynamic models, using methods and techniques of Informatics; these models are then brought to life on a computer. In this way scientists are able to model the reality of their discipline in "dynamic symbols" [2], a virtual dynamic reality.

The study of an informational discipline in higher education typically consists of three parts - elements of study in the discipline itself, elements of Application Oriented Informatics and elements in which discipline and informatics are integrated. Application Oriented Informatics elements include modelling methods and techniques, design of data structures, programming methods and techniques, programming in a discipline specific high programming language, software engineering, and methods and techniques for human-computer interfacing. Shaw [3] mentions two examples of informational disciplines and refers to programmes at Carnegie-Mellon University, George Washington University, the University of Illinois, the University of Colorado in the U.S. and the University of Toronto in Canada. Other examples are also found at Nijmegen University (NL) in the areas of organisational and business science, the languages, psychology, the exact sciences and medicine [1]. At the IFIP working conference "Visualisation in scientific computing: Uses in university education" Weber [4] also presented an example in Chemistry.

INFORMATICS AND ITS APPLICATIONS IN SOCIETY

In industrial countries, society is rapidly changing from a supply driven society to a demand driven one. At the same time the organisational framework is changing from complex hierarchies for mass production to flexible, demand driven network organisations. In these organisations people work as responsible and capable individuals in multi-disciplinary teams instead of as anonymous parts in a Tayloristic hierarchy with its strong functional concentration. In the development of applications of Informatics in society, phases may be discerned of which the most recent offers a strong technology support to the new forms of organisation and work [5], [6].

Automation: empowering the process
In the first phase of development information technology was used to automate simple, standard "white collar" processes, such as business accounting.

Information: empowering the individual
The miniaturisation of computers lead to development of tools supporting the professional. Instead of being automated, the personal work processes were enhanced by this integrated support, thus empowering the individual to handle far more complicated and responsible tasks.

Communication: empowering the individual in the process
Currently powerful personal computers are integrated in local area, wide area and global networks. We are at the beginning of developments which will bring us computers as personal, intelligent agents in communication networks interacting on our behalf with other such agents, and humans. The new organisational structures are in need of this type of technological support and draw heavily on these new technological developments, bringing their introduction and integration into society quickly forward.

INFORMATICS AND ITS APPLICATIONS IN SECONDARY EDUCATION

The uses of information technology and the study of the underlying scientific discipline Informatics in secondary education only to a certain extent have followed developments in society, regardless of whether the schools are situated in developed or developing countries.

Automation: empowering the teaching process, learning about automation
In the first development phase, automation, computers were used to automate clerical teaching tasks and an ill-fated effort was made to automate the teacher. In learning the subject of informatics or computer science dealt with the writing of simple programs in a computer language and with topics like the organisation of a computer centre.

Information: computer literacy and applied informatics
In the beginning of the second phase, students in schools learn to apply information technology in computer literacy classes. In a later stage, information technology is also used sparingly in other subject areas. In contrast to other organisations in society, schools are not able for the most part to empower individual students with a learning environment into which information technology is integrated. An overview over the problems may be found in [7]. The learning of informatics, that is learning of principles, methods and techniques of the science of informatics, mostly takes the form of learning to program in a higher programming language. An overview over developments here may be found in [8].

Communication: Empowerment of the individual student in the learning process?

In general developments in schools have become stuck in the second development phase. The integration of information technology into the learning process, thus empowering the individual student proves impossible because of traditional constraints of content and organisation. However, under pressure of the demands of a society entering the phase of communication, education will follow developments in society [6].

DEMANDS ON EDUCATION

Other student competencies

Developments in society bring forward a strong need for a workforce which is capable of high-level work in multi-disciplinary teams supported by integrated information technology [5], [6], [7]. This workforce has to cope with the increasing dynamic complexity of the working processes in society. For this they need higher order modelling capabilities for handling and interpreting dynamic models of reality; they need to be able to handle and interpret a virtual reality [2]. Secondary education will have to deal with these needs. It is now far less important to learn rote skills, such as reading, writing and arithmetic, then it is to learn higher intellectual skills, such as analysing, abstracting and modelling. And students have to learn to work in multi-disciplinary teams supported by integrated information technology [9]. That societies moving through constant technological change are in demand of people with high-level skills, higher than ever before, is also clearly stated by the Organisation for European Co-operation and Development [10]:

> "In the manufacturing sector, requirements are shifting towards multi-skilled roles, teamwork and conceptual skills. In the service sector the focus is much the same, but with additional emphasis on customer and communication skills."

Conceptual and communication skills are central skills in an increasingly knowledge-intensive economy. Secondary education should allow students to develop such high-level skills.

Modelling of dynamic processes

To be able to cope with the increasingly complex dynamics of a world consisting of communicating processes, students need high-level modelling capabilities allowing them to build dynamic models of reality

[2]. In higher education such needs are already addressed in informational disciplines. In these disciplines students learn to create dynamic models of reality which are implemented in special-purpose, conceptual "programming" languages which are specifically developed to model special micro-worlds of reality. This type of discipline integrated informatics modelling is expected to also penetrate into secondary education [9].

Integration of IT competence

Parents in society have experienced individual empowerment by information technology integrated into their working tasks. They will put a demand on education to use the same integrated technology to prepare their children for that society. This gives strong support to the integration of IT competence into the set of competencies of the educated person [7], of which competence another style of working and studying will an integral part.

Multi-disciplinary education in teams

More and more people in society will have to work as a responsible individual in a team tackling tasks which need a variety of skills to perform. Collaborative learning in multi-disciplinary teams may therefore be expected to be introduced into education. For developing countries this offers the extra advantage of maximising the use of scare resources. The organisation of education will need to change from a complex hierarchy suited to mass education to a flexible, student driven communication learning network. Technological developments will both support and demand this change [6].

INTEGRATION OF INFORMATICS IN THE SECONDARY CURRICULUM

Society is changing its structures and this change is supported by information technology. There is a demand on education to develop capabilities in the students which will allow them to cope in this changing society. This demand will lead to an integration of informatics into other disciplines or subject areas which integration is aimed at the development of high level modelling skills. Demands from society will also lead to real integration of information technology into the learning process and a change in content and organisation.

Discipline integrated informatics: a new literacy

Ershov in his keynote address at the IFIP World Conference on Computers in Education in 1981 [11] pointed at the emergence of a new literacy as compared with traditional literacy (reading and writing); in his view programming was the key feature of this literacy. This metaphor has been taken further [2] to describe a new literacy emerging which is based on the capability to construct executable symbols (programming in a generalised sense) reflecting complex, dynamic, conceptual models of reality. Interpretation of such models will allow people to cope with a dynamically changing, complex reality of interacting processes.

This new literacy is built on application oriented informatics and forms, as discipline integrated informatics, part of informational disciplines. Just as in higher education, these informational disciplines will develop in secondary education within the usual subject areas, allowing students to build up higher order modelling skills in that area. Discipline integrated informatics will be the new "informatics for all", aimed at development and interpretation of dynamic models. This "informatics for all" will be not only an introductory course, as these skills will have to be mastered by all the students in order to prepare them for society. Some indication of what is to come may be found in the UNESCO Curriculum for Schools [12], where a module on applications of modelling is introduced based on discipline integrated "Subject Oriented Programming" with tools, such as a spreadsheet or a computer algebra system.

Integrative use of IT in multi-disciplinary teams

Just as in society, students will learn to use information technology in an integrative way, empowering them as individuals in the context of multi-disciplinary teams. The integrated use of information technology will force subjects to change in content, just as this use changes the content of work in business and industry. A clear example may be found in mathematics where, after the changes brought about by the hand-held calculator, the use of computer algebra tools like Maple, Derive or Mathematica now deeply influences the work of professionals. This influence will inevitably lead to changes in the content of mathematics as a subject in secondary education.

Application oriented informatics

Application oriented informatics is part of the new literacy in secondary education, aimed at all students. In the teaching of application oriented informatics there will be a move away from learning to program in traditional programming languages. Instead, students in secondary education will learn to program in conceptual languages which model "micro-worlds" of interest to the students or to a particular discipline. The study of application oriented informatics will include: modelling methods and techniques, the design of conceptual data structures and program design in high level, conceptual programming languages.

"Informatics for a few" will also be offered in preparation for particular jobs or particular further studies. In informatics for a few some more advanced topics such as software engineering and design of human-computer interfaces will be included.

CONCLUSION

Developments in secondary education with respect to informatics and its applications may be characterised according to the phase of technological development. With each phase in development (automation, information or communication) the integration of informatics and its applications in education is further developed.

From the computer programming and social aspects of computing of the *automation* phase emerged computer literacy, ad-hoc use of application tools in subjects, and some elements from application oriented informatics, i.e. modelling and general purpose programming, of the *information* phase. By the time of the *communication phase,* discipline integrated informatics will emerge as a new literacy and there will also be information technology literacy, integrated in the subject areas. The content of subject areas will change, just as educational organisation and the way students learn. There will also be application oriented informatics as an elective, dealing with modelling, conceptual programming and data structures, human-computer interfacing and software engineering.

REFERENCES

1. van Weert, T. J. (1992) *Application Oriented Informatics and Informational Disciplines: A symbiosis bridging the gap*, in: R. Aiken (ed.), Information Processing **92** (II). Elsevier Science Publishers B. V., Amsterdam, 144-150.

2. van Weert, T. J. (1988) *Literacy in the Information Age,* in: Sendov, Bl. & Stanchev, I. (eds.), Children in the Information Age, Pergamon Press, Oxford, 109-122.

3. Shaw, M. (1991) *Informatics for a new century: computing education for the 1990's*, special issue of Education & Computing, 7 9-17.

4. Weber, J. (1994) *Visualising microscopic molecular worlds in chemical education,* in: Franklin, S. & Stubberud, A. (eds.), Visualisation in scientific computing: Uses in university education, Elsevier Science Publishers B. V., Amsterdam (in print).

5. Hammer M. & Champy, J. (1993) *Re-engineering the corporation.* Nicholas Brealy Publishing, London, 83 - 101.

6. van Weert, T. J. (1992) *Informatics and the organization of education,* in: Samways, B. & van Weert, T. J. (eds.), The impacts of informatics on the organization of education. Elsevier Science Publishers B. V., Amsterdam, 15-24.

7. van Weert, T. J. (1994) *Education and computers; Who is in control?,* in: xxxxxx Information Processing 94, Elsevier Science Publishers B. V., Amsterdam (in preparation).

8. Taylor, H. G., Aiken, R. M. & van Weert, T. J. (1991) *Informatics education in secondary schools,* IFIP Working Group 3.1, Guidelines for Good Practice. IFIP, Geneva.

9. Ruíz i Tarrago, F. R. (1993) *Integration of Information Technology into Secondary Education: Main issues and perspectives,* in:
van Weert, T. J. (ed.), IFIP Working Group 3.1, Guidelines for Good Practice. IFIP, Geneva, 7-15.

10. OECD (1988) *New technologies in the 1990's - A socio-economic strategy,* OECD, Paris.

11. Ershov, A. P. (1981) *Programming: the second literacy,* in:
Lewis, R. & Tagg, D.(eds.), Computers and Education, (North-Holland Publ. Co., Amsterdam, p. 1-7.

12. J. D. Tinsley & T. J. van Weert, eds., (1994) *Informatics for secondary education, A curriculum for schools.* UNESCO, Paris.

Tom van Weert is dean of the undergraduate School of Informatics (Computer Science) of the Faculty of Mathematics and Informatics of the University of Nijmegen. He teaches management of large software projects to students developing real software applications in multi-disciplinary teams. Previously he worked in secondary teacher education, teaching mathematics and informatics, and prior to that as a computer system engineer. He has been active within IFIP Working Groups 3.1 on secondary education, 3.2 on university education and 3.5 on vocational education.

13

Promoting interdisciplinary and intercultural intentions through the history of informatics

Klaus-D. Graf

Institut für Informatik, Freie Universität Berlin
Berlin, Germany

ABSTRACT

The integration of informatics and information technology into schools should not only be oriented towards subject matter and methods but also towards general current demands of society on education. Interdisciplinary and intercultural knowledge and abilities form part of these demands. They can be promoted by aspects of the history of informatics such as early binary coding in China or early calculating machines emanating from roots in different cultures - Arabian, Chinese and European.

Keywords: culture, innovation, instruction, interdisciplinary, pedagogy

INTRODUCTION

The integration of new subject matter and methods in education at our schools is normally influenced by different intentions, very often contradictory, one more passive and one more active or even aggressive. The emphasis of the first lies in the support or the extension of existing matter and methods; the emphasis of the second lies in the introduction of new subject-matter and methods, partially accompanied by ousting traditional matter. Different opinions also occur about the character of innovative intentions. They can be strongly oriented towards the subject or strongly determined by the general expectations of society from education.

These very general statements relate directly to issues associated with the introduction of elements from informatics or information technology in education. The fundamental position of this paper is that the main task is not the extensive integration of informatics and information technology in education, but the improvement of education in general, as expected and demanded by a society faced with ever-growing problems of enormous complexity. Informatics will be of great importance for this task, and not just related to the more narrow objectives of its own.

INFORMATION TECHNOLOGY AND SOCIETY

Ruiz reports that decision makers in education agree that modern societies rely increasingly on larger numbers of individuals with high-level knowledge and skills at their disposal [1]. This means, in addition to elementary cultural techniques such as reading, writing and calculating, school has to teach other skills such as analysing, abstracting and modelling. Conceptual and communication skills are also required in the fields of production and service. An increasing range of jobs demand abilities and talents from individuals who know about their responsibilities in a wide context, and many of these people must be able to work in multi-disciplinary teams.

Moreover, further high-level skills than those listed above are needed if one considers the challenges to societies by universal problems such as the world economy or environment. These problems are of extreme complexity and require a great amount of international interaction. Any success in these activities is dependent on mutual understanding and acknowledgement of different traditions and attitudes concerning elements of these problems, such as mass production, environmental protection, political and social structures. This understanding is required not only from a few experts at a high level, but from anyone engaged in problems related to foreign concerns. Readiness and goodwill have to be developed, so more attention has to be given to intercultural intentions in education. Teaching foreign languages is only a beginning; it must be accompanied by reflecting the relations between any two respective nations and their cultures in present time and in history. This in turn leads to a need to understand the influence of foreign culture on other cultures. These are demands on a very high level; it is well understood that in school only very elementary first steps can be taken. But they have to be taken there to ensure a modest contribution of our school education to solving the problems of our societies.

At the present time, there appear to be substantial problems in relationships between Arabia, China and the Western World. More mutual knowledge about the history of these cultures could be a small contribution to tackling these problems of global relationships in a peaceful way.

THE ROLE OF INFORMATION TECHNOLOGY

There are avenues in most subjects in school to pursue intentions such as the ones discussed above and every subject could make its contribution. In particular, informatics and information technology form a rich source for contributing to such mutual understanding.

Analysing, abstracting, modelling and conceptual skills are fundamental ideas in software engineering, and elementary techniques from this field can be integrated into informatics education. Communicational and inter-disciplinary skills are developed when solving problems by project teams. A feeling of responsibility for one's work can be developed when the consequences of informatics applications are considered systematically as a constituent part of the project and the project team. Also, intercultural awareness can be developed through studying the history of informatics; I will address this point further on in this paper.

Didactical model

I comment elsewhere on this practice [2]. Below is a representation of a didactical model for the teaching of informatics and its implications, which can enforce the consideration of intentions such as the ones related to responsibility, interdisciplinary cooperation or intercultural context as well as subject matter and methods of informatics. Instead of just putting a problem from informatics or information technology in the classroom and asking how to solve it, this problem should be regarded in a string of three questions.

WHERE FROM ⇒ WHERE HERE AND NOW ⇒ WHERE TO?

- *Where from?*: Where does the problem come from? What are the causes or motives for it? Who is interested in the solution and who helps or opposes?
- *Where here and now?*: Which resources do we have here and now to solve the problem? How can it be done?
- *Where to?*: Where will the solution of the problem lead us to? Will we get what we want, and what else? What are the impacts and side effects in our environment, in society?

We can extend this scheme to questions to put to informatics and information technology in general.

Where does informatics come *from?* To answer this question we study the motives and causes of people who invented informatics tools or methods, we study the cultures or subcultures who were interested, we study the disciplines from which elements of informatics grew.

Where is informatics *here and now?* What is informatics? To answer this we study tools for description, analysis and design of information processing.

Where does informatics lead *to?* To answer this we study the social and economic impacts of information technology, including impacts on law. For this we need an objective technique of criticism and scientific research about consequences of technology.

THE HISTORY OF INFORMATION PROCESSING

Some key events

The history of information processing, information technology, computer science and informatics also furnishes us with striking evidence of the importance of intercultural events and activities, leading to problem solving in the international scientific world as well as in other fields such as the international economy and traffic control.

There are a number of key events, with particular intercultural and interdisciplinary aspects, which gave rise to important technical and social developments in different societies and cultures [3].

Numbers: Number systems and calculation rules were developed from about 2000 B.C. in Babylon (today Iraq) and Egypt to be used in public administration, trade, mathematics, astronomy, calendars, giving rise to exact technics and sciences.

Scripts with alphabets: At about 1300 B.C. the Phoenicians, living in today's Syria, developed a set of phonetic characters, semantically meaningless (today called an alphabet), for an efficient way of writing. Their motive was the easy documentation of contracts, laws and rules in different languages. This method spread quickly to the ancient Greek and Roman worlds; the effect was reliability in storage and transport of information.

Algorithms: At about 200 B.C. in India, the invention of the decimal point system, using the number 0, was another key event. This technique was transferred to Arabia in the 9th century. In Bagdad Alᵛ Hwarizm then developed his "algorithms", highly appreciated in both trade and astronomy. Together with the decimal system, they became popular in commerce and trade first in Italy in the 15th century and finally in all Europe.

Dyadic numbers: In 1679 in Germany, Gottfried W. Leibniz developed the dyadic system for the denotation of numbers, his motives being basically mathematical but also philosophical. Joachim Bouvet, a French Jesuit in Beijing, China, at the court of Emperor Kang Xi, compared the dyadic numbers to the hexagrams, which were written down 5000 years before in China. This caused deep reflection on Leibniz' part also about the religious and philosophical importance of the binary elements, the digits 0 and 1, or the symbols yin and yang, -- -- and ----.

At the same time, the importance of the dyadic system for calculating machines became apparent to Leibniz. He even described a machine, operating with input from this system. The idea of such a machine vanished in history, however, and it was not until 1938 that it was recreated by Konrad Zuse in Berlin, Germany when he constructed the first modern computer working with dyadic numbers.

Calculating machines: Algorithms are symbolic machines [4]. After they had been established, the search started for real machines as processors for algorithms. In 1623, Wilhelm Schickard from Tübingen, Germany, described and had constructed a calculating machine, which mechanised addition and subtraction as well as some steps in multiplication and division. In our context three factors are significant: Schickard's motive was to support astronomical calculations done by Joachim Kepler at this time; he integrated the technique of Napier's bones from England (1617); and he relied on the well-developed art of clockmakers, even calling his machine "Rechenuhr" - calculating clock. His machines were destroyed and forgotten; Schickard's idea was only rediscovered in this century from some of his messages and sketches.

In 1642 Blaise Pascal in Paris, France presented a calculating machine for addition and subtraction, which became well-known in his time and which caused the evolution of more machines. Leibniz was also motivated by Pascal's invention and from 1672 on he worked on concepts for the first machine to perform all four species of calculation completely automatically, which was finished in 1694.

Automata: Another key event can be found in Ancient Greece at about 100 B.C., when Heron of Alexandria invented automata for different control problems, serving religious purposes such as opening and closing temple doors. From our point of view these were programmed machines, as were the "androids", which became popular in Europe in the 17th century. These could perform activities such as playing an instrument or writing with a pen, following a mechanically stored program. These machines were made for entertainment only, although there is some deep reaching philosophical relation to the problem of an artificial human, created with biochemical or technical means.

More pragmatic motives are behind the first attempts to control mechanical weaving machines by punched cards (Falcon 1728, Jacquard 1805 in France) or to mechanise the evaluation of public or commercial statistical data from punched cards (Hollerith 1887, U.S.A.).

Universal machines: Programming and calculating were united in Charles Babbage's design of an Analytical Engine, taking up Jacquard's punched cards, and expected to do all kinds of calculations (Great Britain, 1836).

Thus the idea of a universal computer became more concrete, an idea which had already been anticipated by Leibniz. In today's view he was thinking of a symbolical machine to solve general logical problems, a "calculus rationcinator", universal calculus of thinking. At the same time he discussed a real machine, a "machina ratiocinatrix", which should mechanise the logical processing.

Alan Turing (Great Britain, 1941) wanted to find out which problems are accessible by such a machine. His answer was another symbolic machine, his universal automaton, better known today as the Turing machine . With this he laid the theoretical foundations for the automation of symbolic machines. All our modern computers are models of this concept.

Many more key events like the ones mentioned above can be traced in the history of informatics after Babbage or after Zuse and Mauchly/Eckert, which allow exploitation in education through an informatics model which asks "Where from, where here and now and where to?".

I cannot go into all of these in this paper. Instead, two examples of subject matter for informatics education shall be presented in some more detail. They may look rather exotic and marginal in the beginning. But looking closer they reveal many elements satisfying subject related intentions as well as general intentions of informatics education.

Hexagrams from the I Ching

The 64 hexagrams shown in the centre of figure 1, taken from [5], form a complete binary code of length six. Writing 0 for the broken lines (ying) and 1 for the solid lines (yang), from left to right instead from bottom to top as in the hexagrams, you get the system of numbers from 0 to 63 in dyadic (dual) representation.

These hexagrams became well-known through their role in the Chinese 'I Ching', the Book of Changes, where they represent 64 mental states of nature, society or individuals. Changes in their structure by exchanging yin and yang lines characterise the eternal transitions between these states. The hexagrams are older than the 'I Ching', at least 4000 years old. They were possibly composed as pairs of the eight trigrams shown in figure 2, taken from [5], which go back to the Chinese sage Fu Hsi.

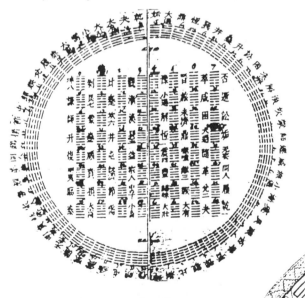

Figure 1: 64 hexagrams forming a
complete binary code [5]

Figure 2: Eight trigrams [5]

Leibniz, who (through figure 1) in 1701 learnt about the hexagrams from Bouvet, a Jesuit at the Court in Beijing, called them "the eldest monument of science". He was fascinated by their relation with the dual numbers which he had just made public. He started a discussion with Bouvet whether the old Chinese had known about the numerical importance of the hexagrams, and if the contemporarians had just forgotten about this.

It is very likely that the old ones did not know. Their intentions with the hexagrams were very different from those of Leibniz. They were looking for a systematic description of the universe. One attempt postulates that at "the very beginning" (Tai Chi), two opposing elementary powers were created, named yang and yin later. These gave rise to four states - the four seasons - and these again to eight trigrams - representing, among other interpretations, eight elementary phenomena from nature, shown in figure 3. This figure is nothing but a binary coding tree. If you expand it in three more steps, you arrive at the 64 hexagrams. Their meanings again are of natural or philosophical character, and so their ordering in the I Ching is different from the one suggested by the binary tree. It is very likely that the ordering in the I Ching refers rather to the changes which can occur from one state to another. At the same time the arrangement of the yins and yangs in each one of the hexagrams has no arithmetic meaning; it stands for the interaction of the yin and yang powers.

The mathematical ingenuity expressed in this binary code is also underlined by the Chinese document from the 12th century, or possibly earlier, shown in figure 4, taken from [6]. Actually it shows a surprising graphical representation of a complete binary tree.

There exists another ordering of the hexagrams, contained in Richard Wilhelm's translation of the I Ching. It has a clearly combinatorial character, starting with the "heaven-hexagram" with six yangs, i. e. zero yins, followed by six hexagrams with exactly one yin, 15 hexagrams with two yins, 20 with three yins and so on.

It is interesting to remark that research done at Rutgers University in New Jersey with American first graders showed that they prefer this kind of ordering when the following problem is put to them: "Here are a lot of blue cubes and red cubes. Please construct all different piles of 3 (or 4, or 5) cubes".

Another proof for the combinatorial skills of the old Chinese scientists is a table from the 10th century, showing the 81 tetragrams, which you gain from three elements, ------, -- -- and -- -- --, for example.

Chinese calculating machines from the 17th century.
Some calculating machines were constructed in the Imperial Palace in Beijing at the end of the 17th or early in the 18th century, or they were carried there as gifts from Europe. Two of these machines were rediscovered in 1962, 8 more in 1978. An article about these was published in Chinese in 1980. Two machines were shown in Brussels in 1988. I was able to see three of them in Beijing in 1991. In 1992 Michael Williams published a report about the Chinese article in the Annals of the History of Computing, after discussion with the authors, Bai Shangshu and Li Di [7]. A recent interpretation is given by myself in another paper[8].

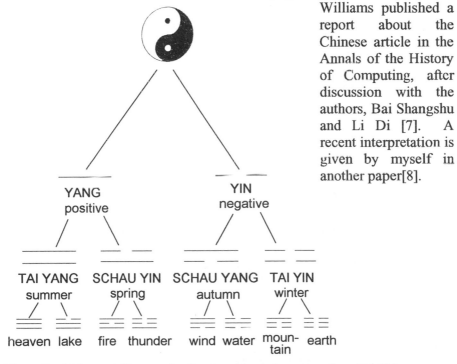

Figure 3: A binary coding tree for for elementary phenomena from Tai Chi

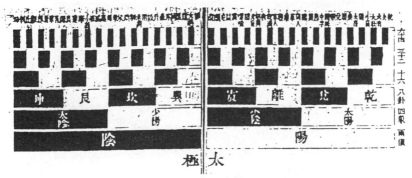

Figure 4: 12th Century representation of a complete binary tree [6]

Six machines work with decimal tooth-wheels including devices for the carries (figure 5). They allow addition and subtraction. The construction possibly goes back to Christian Huyghens, who developed it after studying Pascal's machine (1643). The technique of the wheel operations is different from Pascal's (figure 6).

Four more devices, supporting multiplications, rely on cylinders each carrying 12 of Napier's bones. They are similar to "calculating cylinders" made by Caspar Schott (1664) and to the multiplication section of Schickard's machine, but they also show impressive improvements by the Chinese constructors.

It is very likely that information about the European devices came to Beijing via the Jesuits, who stayed at the Chinese Court during the reign of Emperor Kang Xi. There exists a letter from Caspar Schott to one of them. It is unknown, however, where exactly the information about the wheel machines came from.

Figure 5: Disc calculating machine with decimal tooth wheels
(courtesy of Professor Bai, Beijing)

Figure 6: Internal view of disc calculating machine
(courtesy of Professor Bai, Beijing)

CONCLUSIONS

It is hoped that these two examples illustrate how a teacher can arrange teaching sequences in informatics education which contribute to the development of inter-cultural studies. A positive advantage may be gained for informatics by putting more weight on coding and coding algorithms, for example, instead of on arithmetical problems, or by learning about calculation machines by comparing different historical models. The cross-cultural influences on early developments in informatics is clear, and might be used to substantial effect within informatics education today.

REFERENCES

1. Ruiz i Tarrago, F. R. (1993) *Integration of Information Technology into Secondary Education: Main Issues and Perspectives.* In: van Weert (ed.) Guidelines for Good Practice, IFIP Working Group 3.1, Geneva .

2. Graf, K. D. (1991) *Powerful Means in Mathematics and Computer Science Education: Mathematical, Logical, Mechanical and other Roots of Computer Science in History.* Journal of the Cultural History of Mathematics 1. Mathematics Education Society of Japan.

3. Rode, H. & Hansen, K.-H. (1992) *Die Erfindung der universellen Maschine,* Hannover.

4. Krämer, S. (1988) Symbolische Maschinen, Darmstadt.

5. Zacher, H. J. (1973) *Die Huuptschriften zur Dyadik von G. W. Leibniz,* Frankfurt am Main.

6. Needham, J. (1954-1965) Science and Civilisation in China, I - IV, (1+2), Cambridge.

7. Bai Shangshu, Li Di & Williams, R. M. (1992) *Chinese Calculators Made During the Kangxi Reign in the Quing Dynasty.* IEEE Annals of the History of Computing **14** (4).

8. Graf, K.-D. (1994) *Calculating Machines in China and Europe in the 17th Century - The Western View.* Brunnstein, K. & Raubold, G. (eds) 13th World Computer Congress '94 Elsevier, North Holland.

Klaus-D.Graf studied mathematics, physics and education at Chemnitz in Germany and at the University of Illinois in the USA, receiving his PhD in mathematics in 1966 from the University of Mainz, Germany. He lectured in cybernetics and became a professor of mathematics education in Neuss in 1971, and a professor at Freie Universität, Berlin in 1975.   From 1980 he has specialised in informatics education, computer use in mathematics and CAL.   He was a visiting professor in the Republic of China in 1987 and has authored and co-authored textbooks on computer science and information technology in schools.

14

Implementation of computers in Malaysian schools: problems and successes

Zoraini Wati Ahas

Faculty of Education
University of Malaya, Malaysia

ABSTRACT

In this paper, problems and successes that have emerged during an experimental Computer-in-Education (CIE) project in Malaysia are highlighted. The experimental CIE project is taking place in 60 rural secondary schools. The main thrust is computer literacy. The paper discusses problems associated with in-service training, movement of teachers, scarcity of resources and hardware and successes achieved so far.

Keywords: developing countries, government, information technology, literacy

INTRODUCTION

Malaysia's vision is to become a fully industrialized country by 2020. To achieve this vision (Vision 2020), the country needs to turn the present school generation into the leaders of tomorrow. It is essential that people must have information technology skills - defined as familiarity with the use of computers, knowing how to use the computer and knowing how to take full advantage of the computer in various situations.

The Prime Minister stated in 1991 that, in the information age, Malaysian society must be information rich and computer literate if it wants to progress and develop [1]. One of the immediate tasks of the ministry of education is thus to ensure that its students are computer literate by the time they leave school. Since a seven year-old and thirteen-year-old today will be respectively 33 and 39 years old in 2020, they will be then among the main contributing members of the society.

COMPUTER-IN-EDUCATION PROJECT

The Computer-in-education (CIE) project is Malaysia's second attempt to introduce information technology into schools on a national scale. The CIE project is expected to be implemented in stages; it is now in its experimental phase. It was launched in July 1992 by the Minister of Education in 60 rural secondary schools chosen among over 1,400 schools.

Malaysia's CIE project is probably unique because of its implementation in schools in remote areas of the country. The schools are far from the nearest small town, usually surrounded by rubber estates, oil palm estates, paddy fields or rain forests. They are often "in the middle of nowhere". The students come from poor families whose fathers are farmers, rubber tappers or work on the estates. Their mothers are usually housewifes and both parents are poorly educated.

To illustrate how remote these schools are, four of the schools do not have telephones, one is only accessible by boat, about eight schools are served by dirt roads and are at least two to four hours bumpy drive away from the nearest small town. To get to some of the schools from the city, one would have to take a small plane, usually a twin engined Otter, to a small clearing in the jungle and get a taxi. The taxi ride is usually another two to four hours.

Rural schools were chosen over urban schools to correct the existing imbalance between the schools that already have and schools that do not have computers. More than 800 other schools have computer clubs. The challenge to introduce computers into these schools is thus very great.

The ministry formulated the topics for a computer literacy syllabus, trained teachers and supplied hardware and software to schools. The total cost to implement the CIE project in the 60 schools was RM 4.8 million (US 1.8 million). Computer Literacy is taught to 13 and 14-year olds, that is, those in Form One and Form Two for two years. Each school received 20 made-in-Malaysia PCs that are networked to a 386-SX server (16 MHz,

4MB RAM and 120MB hard disk, trackball), an Electronic Imaging System (EIS) and a printer.   Each computer is shared between two students and sometimes three depending on the total class enrollment. The software purchased for the schools included WordPerfect 5.1, DrawPerfect 1.1, dBaseIV, Lotus 1-2-3 ver. 2.3, Power BASIC, EJA-BMTeks (a local spell checker), DR-DOS 6.0 and D-Link's LANsmart.

According to a survey among all computer teachers focussing on the early stage of implementation, initial teething problems made teaching difficult [2].   Only about 40 percent of the computer lessons were successfully carried out.

Topics in the Computer Literacy Syllabus for the CIE Project are:

Introduction to the computer	Graphics
Types of computer systems	Spreadsheet
Computer system and operating system	Database management system
	Introduction to programming
How computer processes data	Effects of computer use
Computer ability	Effects on lifestyle
Microprocessor	Computer misuse and abuse
Application packages	Computers in the future
Word-processing	Computer careers

The other 60 percent were not carried out mainly due to lack of teacher confidence as well as hardware and software problems.   However, after one and a half years of implementation many computer classes are now running well.   Nevertheless, problems such as computers breaking down or the EIS not working are still common.

Problems
Implementation of a large project is never easy.   In the CIE project, most of the problems are associated with ineffective in-service courses, movement of teachers, scarcity of resources and hardware.

Teacher selection and training
Two teachers from each school were identified to become computer teachers.   They were chosen from among those most likely to stay in the same school over the next few years to ensure the CIE project is successfully implemented.   In addition, each of the thirteen State Education Departments sent an officer to act as a resource person and to help the CIE Unit co-ordinate the project at state level.   They then

attended in-service courses organized by the CIE Unit of the ministry that is in charge of implementing the project.

Male teachers formed the majority and about half of the teachers were either science or mathematics teachers. Most (58.1 percent) had more than four years of teaching experience and 94.3 percent of the selected teachers had attended at least one in-service course [2]. The first in-service computer course was held in December 1990 and the fourth and last was in July and August 1992. The courses covered an introduction to computer hardware and how to use the respective application package including how to program in BASIC.

The courses were conducted mostly in English by personnel from the software distributor companies. English is the main language used within the private sector but the medium of instruction in schools, institutions of higher learning and the official language of the government is Malay. The teachers in the group are more well-versed in the Malay language. According to several teachers, not only was there a language problem, but the pace was too fast and the approach used was ineffective. Two or more teachers were sharing a computer and this was unsuitable because of their computer inexperience. Furthermore, there were no assessments or evaluations to see how effective the in-service courses were.

Participants felt that the courses were inadequate, particularly for those handling computers for the first time. The problem was augmented when they were unable to continue their practice of the software due to the absence of computers in schools. The computers for the project were not delivered to schools until mid-1992. Thus teachers had very little time to learn on their own. Many teachers felt that it was not easy to teach the computer literacy subject, especially when they themselves were still learning. The most difficult was BASIC programming. Several teachers mentioned they had to leave out this topic.

Perhaps, Malaysia should consider giving PCs to the teachers to bring home after the course. Some of the more successful teacher training programmes in the US have found this to be effective because teachers can use the computer whenever they wish, have more time to explore, take risks and become proficient in the use of the software [3].

Command of the computer syllabus
When teachers were asked how they rated themselves in their knowledge of the more theoretical topics in the computer course, it was found that their understanding was also below the minimum level required.

Among all the software, WordPerfect was the easiest to learn. It was mastered by 56.2 percent of the teachers. Next was DrawPerfect (48.6

percent), Lotus 1-2-3 (42.8 percent), DR-DOS (24.2 percent), dBaseIV (21.4 percent) and EJA-BM Teks (19.6 percent). Lowest on the list was BASIC Programming with only 13.6 percent of the teachers reporting mastery. These statistics were a result of a survey conducted among the teachers in March 1993, eight months after the launch of the CIE project [4]. These figures are not encouraging. It goes against the recommendation by Flake, McClintock & Turner [5] that the teacher should at least know what the student should know. The teachers should be 100 percent proficient before being able to teach.

The percentage of teachers with good or very good understanding of the theoretical topics in the Computer Literacy syllabus is shown below:

Introduction to the computer	53.4%
Types of computer systems	43.8%
Computer and operating system	40.0%
How computer processes data	35.3%
Computer ability	36

Teacher movement

The identified project teachers should have been among those most likely to stay on in the same school. However, as many as one-third of the original group were transferred to other schools. New teachers were appointed to replace them but this posed a dilemma. These new teachers needed time to become involved with the computer classes due to their low level of computer literacy. Basic skills such as copying files from one disk to another and managing the hard disk was a major problem for such teachers. Computer viruses posed an even greater difficulty. Most of the time, teaching was limited to what the teacher knew, thus covering only part of the syllabus.

Hardware

In addition, hardware sometimes caused problems. Electrical surges frequently damaged the computers in some schools prone to attacks of lightning. The server was rather slow and downloading of software took too much time and precious minutes were lost before a teacher could begin teaching. Other problems included hard disk management. Often, in schools with inexperienced teachers, hard disks were corrupted or some essential files lost. In such instances, the teacher resorted to using the computers as stand-alone machines. About 20 schools frequently experienced computer break downs because of serious electrical fluctuations and needed auto-voltage regulators.

Resources
Even more difficult was the scarcity of resources. Only about 20 of the schools had books, reference materials and teaching aids for students in the first year of implementation. The ministry only supplied software packages. Schools did not receive a budget to purchase other resource materials. Teachers found it difficult to prepare lessons and acquire reference materials. Although there were a good number of local reference materials, most of these were either unknown to the teachers or not easily available in the rural areas. Many imported reference materials were available, but were only found in city bookshops. Even if they were easily accessible, the prices would have been prohibitive.

Successes
Overall, students were enthusiastic, looked forward to computer classes and were disappointed when classes are cancelled. This in itself has been a major accomplishment of the project. Without the project, these students would have never experienced computers first hand in the school. An interview with a few Form Two students revealed that they and all of their classmates enjoyed the computer classes (70 minutes per week) and that it was their best-liked subject in school. They felt extremely fortunate to be among the few in the country to be taught computers as a subject. This was further supported by their teacher who stated that when computer classes were held outside normal school hours, the students had no complaints even though it meant coming to school four hours earlier in order to catch the only bus. This was due to time-tabling problems; not all computer classes could be scheduled in the afternoon along with the other subjects. Thus, half of the computer classes were in the morning, outside their normal school timetable.

One particular school located amidst rubber plantations, two hours away from Kuala Lumpur, was particularly successful; here the computer teachers were very motivated about the project. In this school there are 23 Form One and Form Two classes and each class was taught 70 minutes of computers each week. Two computer teachers would normally be able to handle these two classes between them. However, these two computer teachers, like all the other computer teachers in all the other schools, had two other subjects to teach. The other two subjects were considered more important because of examinations held at the national level. The teachers were overloaded in trying to teach all three subjects. To overcome this problem, they trained 17 of their colleagues and today, this school has 19 teachers who are able to teach the computer course. This has been a major accomplishment.

Headmasters have had a significant role in ensuring the success of the CIE project. It was found that schools with headmasters who frequently discussed problems with the computer teachers had more success in implementing the project. The principal of the school referred to above has been very supportive, in spite of not being a computer user himself. He has even managed to get air-conditioners for the computer room, a luxury that one rarely finds in any Malaysian school.

CONCLUSION

Successful implementation of any project, particularly one that is this large, with schools scattered throughout the country and in remote areas, is extremely difficult. Some of the project schools do not have telephones or are only accessible by boat or dirt roads. In a developing country such as Malaysia, most facilities and resources are concentrated in cities and large towns. The administrative centre is in Kuala Lumpur, which is the capital city and the hub of the nation. Both the CIE Unit and the hardware supplier are located in Kuala Lumpur. However, it is strongly believed that if teachers were well-trained, many of the problems described above would not have occurred. At present, the CIE project lacks effectively-trained teachers both in quality and quantity. To succeed, it has to ensure that its in-service training is effective. The teacher is a key to technological integration, and in addition, this teacher needs strong administrative support and adequate resources.

REFERENCES

1. Mahathir, M. (1991) *Malaysia: The way forward*. KL: Centre for Economic Research & Services, Malaysian Business Council.

2. Sulaiman, S. & Zoraini, W.A. (1994) *Pelaksanaan projek percubaan komputer dalam pendidikan di sekolah menengah (The implementation of the Computer-in-Education experimental project in secondary schools)*. Proceedings of EDUCOMP '94 National Symposium, Penang, Malaysia.

3.   Lare, D. *The teacher as the key to technological innovation.* Proceedings of the Tenth International Conference on Technology and Education, 1 (110-12).

4.   Sulaiman, S. (1994)   *Penilaian pelaksanaan projek percubaan komputer dalam pendidikan di 60 buah sekolah menengah (The evaluation of the implementation of the Computer-in-Education experimental project in 60 secondary schools).* Unpublished MEd. thesis, University of Malaya.

5.   Flake, J.L., McClintock, E.E. & Turner, S. (1990)   *Fundamentals of computer education.* CA: Wordsworth Publishing.

Zoraini Wati Abas is an associate professor who teaches educational computing and instructional technology courses at the university.   She has been chairperson of the Malaysian Council for Computers-in-Education which promotes computer use in education. She has organized three national educational computing symposiums in Malaysia. She has helped implement a computer literacy programme in junior science colleges and sat on several committees looking into computer associated programmes in schools and universities.   She has written several books on computers for use in schools, colleges and universities.

15

Interpreting internal school factors on the educational integration of IT

Jordi Quintana i Albalat
Ferran Ruiz i Tarragó

Generalitat de Catalunya, Departament d'Ensenyament
Barcelona, Spain

ABSTRACT

A qualitative study of four secondary education state maintained schools in Catalonia, with a long tradition of using IT, was carried out following an interpretive methodology. This study sought to identify the main internal factors that impinge directly on the quality of integration of information technology into these schools. These factors appear to be the schools' internal decision making procedures, the basic pedagogical conceptions on the access to IT by the students and its use, and the participation levels of teachers in the innovation process. Drawing on the information gathered, some general conclusions are put forward.

Keywords: case studies, formative evaluation, information technology, integration, organisation

INTRODUCTION

This paper describes a research study into some factors that could contribute to a sound integration of IT in four secondary education schools. The schools of the study were chosen because of their apparent success in implementing IT, success which seemed to be recognized by the teachers themselves and by people that visit or are acquainted with these schools. This success was also often rather vaguely perceived in terms of positive learning outcomes of students gained by using IT, and in terms of user satisfaction, both by students and teachers.

The purpose of the research was to summarize the fragmentary and dispersed indications of such perceived successes on the process of the educational integration of IT. Also, taking into account that evidence does not speak for itself, the aim was to interpret this evidence in its own context, generating knowledge on the internal mechanisms of IT-related innovation at school level.

Rather than a simplistic and reductionist interpretation, the integration of information technology in education should be understood as the everyday and standardized use of computing resources in the widest range of curricular subjects, with a reasonable equilibrium between its use as an instrument for teaching, an instrument for learning and a learning instrument.

THE RESEARCH MODEL

Research model

The so-called naturalist research paradigm (Husen [1], Martínez [2]) provided the setting for this study on the integration of information technology into these schools. Without wanting to go over the conceptual and methodological characteristics and differences between the positive or rationalist and the naturalist or interpretive paradigm (Benedito [3], Blomeyer & Martin [4], Cook & Reichard [5], Elliot [6], Gimeno & Prez [7], Taylor & Bodgan [8] and others) it may be worth commenting on some of the most important factors of the naturalist paradigm.

As Guba [9] points out, the naturalist approach draws near to reality without preconceived theories or hypotheses. The design of the research still develops as the research process itself is actually carried out: it is concerned with an open and resultant investigation model, in which the subjectivity of the instruments used is taken into account and in which maximum importance is attached to relevance. This research model, which is of a qualitative type, is based on the idea that there is not just one reality, but instead, multiple interpretations of facts. This enables the study to be centred on the significances and their possible values in a specific setting or context.

According to Goetz and LeCompte [10], the objective of all ethnographic research is to describe the "characteristics of variables and phenomena, with the aim of generating and perfecting conceptual categories, discovering and validating relationships between phenomena, or comparing the constructions and hypotheses generated, from phenomena observed in different settings". Torres, as quoted in the prologue of Goetz and LeCompte, states that "the aim of educational ethnography is focused on discovering what occurs daily, by means of contributing significant data, in the most descriptive way possible, in order to later interpret and be able to understand that data and intervene more suitably in this ecological niche known as classrooms".

Following this methodology, the research was carried out focussing on three main areas of qualitative analysis of the integration of information technology into these schools: the school's internal decision mechanisms, teacher's conceptions on students and IT, and teacher participation and opinions.

Characteristics of the schools
The four schools in the study were state maintained schools of very different sizes, urban settings and levels. All of them had an important history of using computers, publicly recognized, and they have had links with other schools. In this paper they will be referred to with the numbers 1, 2, 3 and 4. Their main characteristics are given in Table 1 below:

	School 1	School 2	School 3	School 4
Number of students	282	304	970	1,427
Students aged 12 or older	57	94	970	1,427
Number of teachers	17	25	50	97
Number of computers	15	27	16	120
Student/computer ration	18.8	11.3	60.6	11.9
School environment	Affluent urban area	Middle class small town	Middle class middle size town	Middle class middle size town
Average training courses followed by teacher	3	3	0.5	n/a

Table 1: The four study schools

Schools 3 and 4 are just secondary level, whilst 1 and 2 are primary schools which also cover the first two academic levels of secondary education. The internal functioning of schools 1 and 2 strongly relates to the pedagogic approaches and practices of both primary and secondary levels.

Research process
The research process included several stages [11]:

- initial contact with the schools with explanation of the aims of the research, negotiation of the methodology, guarantees of confidentiality, use of the data, and use of the conclusions (stage 1);
- collection and analysis of relevant documentation provided by the schools, with specific attention to the inclusion of references to information technology in the documents and to the level of curricular integration of IT (stage 2);
- carrying out of structured interviews with members of the board, the IT co-ordinator and with teachers (stage 3);
- analysis of the individual interviews (stage 4);
- detailed writing of a provisional report on the findings of each school (stage 5); and, drawing up of the conclusions of the research (stage 6).

Stage 3 was enriched by observations carried out in the computer labs while being used by students, and by informal conversations with teaching staff in school breaks. In these talks information was sought on the degree of satisfaction of teachers concerning the use of IT by the teachers themselves and by pupils. The design of the structured interviews was modified during the research process. In almost every situation, the interviews turned into conversations in which the questions asked and answers given were formed indiscriminately by the researchers and teachers, thus setting up a constructive and reflexive conversation about the quality of information technology integration in the school.

INTERPRETING THE RESULTS

Analysis of the integration of IT
The analysis of this integration was carried out in three main areas which provide information about the different aspects of the schools:

- the procedures of decision making;
- the schools' conceptions about the access of students to IT and the use they make of it; and
- the teachers' participation and opinions.

Schools' internal decision mechanisms

The educational regulations establish that in Catalonia's state maintained schools the school board of directors is elected by the faculty for a given period. In practice there are a variety of possibilities of complying with such regulation: either the school endeavours to find a general consensus (school 1), or there is a tradition of rotative access (schools 2 and 4), or, very often, the school follows the usual mechanisms of candidates standing forward in a voting process. This is the case of school 3, in which, additionally, most of the members of the board belong to the same school department.

When some of the teachers of schools 1 and 2 put forward the suggestion for starting IT-related activities it was considered that this concerned all the teaching staff and the corresponding decisions were agreed by the faculty. This approach is still being followed. In school 3, on the other hand, the initiative came from teachers of a single department and since then the decisions are proposed by each different department and taken by the school board. On account of its structure and number of teachers involved, school 4 set up a specific information technology committee whose proposals are submitted to the powerful school co-ordination committee, a very idiosyncratic and widely representative board of this school.

At present, all the decisions in schools 1 and 2, including those relating to IT, are made at faculty level, independently of who makes the proposal (school board, IT co-ordinator and teachers). In school 3, the science and mathematics department which includes IT-related subjects has a certain degree of autonomy but the decisions are made by the school board. In school 4, the specific IT committee still reports to the co-ordination committee which is responsible for making the decisions.

Conceptions on the students' access to IT

Student's access to IT is a fundamental issue that underpins the school philosophy and practice. Schools 1, 2 and 4 have explicit statements in their school policy documents. It is worth considering the nuances of such statements. In school 1, the access of students to IT is considered as a student's right and the school should provide equal opportunities for everyone. School 2 considers access as an opportunity to enhance learning and commits itself to helping most students take advantage of this opportunity. In school 3 access is regarded as a complimentary curricular activity. In school 4 access is regarded as a student's right but students are obliged to make use of it.

Self access opportunities for students and pupils are minimal in school 3 and permitted during their free periods and at lunchtime in schools 1 and 2. School 4 provides the biggest opportunities: any time of day and Saturday mornings including when there is a class being conducted, provided there are spare computers in the computer room.

Teacher participation and views
The availability of information about the educational use of IT is the base of participation. In school 1, the school board of directors and the IT co-ordinator have complete information and teachers have general information. Teachers of a specific level have full information about that level. In school 2, the IT co-ordinator has complete information, teachers and the school board have general information, and teachers of a specific level have full information about that level. In school 3, each teacher knows what he or she is responsible for while the IT co-ordinator has general information. The school board is poorly informed of the details of the IT use in each level and subject matter. In school 4, each department has complete information about their own activities. The IT co-ordinator and the co-ordination committee are fully informed.

According to the school conceptions on students and IT and also on the level of teacher participation, there are different views of the degree of freedom of the use of IT, both for students and for the teachers themselves. Schools 1 and 2 state that teachers have the choice of using IT and, in school 2, the school climate encourages them to do so. Nevertheless it is accepted that certain teachers scarcely use IT. In school 1, the use of IT is mandatory for all pupils because of the decision taken by the school. In school 3 teachers are completely free to choose but whether or not the department as a whole has decided to use IT will affect their decision. In school 4 the degree of freedom depends on the decisions taken by the co-ordination committee and the school priorities.

CONCLUSIONS AND RECOMMENDATIONS

The research shows that school 3, whose organisation is based on the autonomy of the school departments, lags far behind in the educational integration of IT because of the lack of a global school project. The decisions concerning the use of computers in the curriculum appear to be related to specific people and to specific subjects. The researchers were surprised by the poor student-computer ratio, which can be attributed to the compartmentalization of resources and goals. An unexpected outcome of the research was the beginnings of a reflection process inside the school which could lead to the implementation of a global IT school policy. School 4, which also has a strong departmental structure, avoided this danger from the very beginning by creating a widely representative specific IT commission. Although schools 1 and 2 initiated IT activities in a very personalised way, they quickly moved towards the participation of the faculty by setting a global school policy, whose implementation was eased by their much lower enrolment and by their structure which is not based on the autonomy of school departments.

From the research that was carried out, several recommendations can be deduced. Firstly, the success and the quality of the integration of information technology into education directly depends on the organisation of the school and its versatility, on the internal information channels, on the processes of decision making and on the collective responsibility assumed by the teachers. Specifically, it is worth stressing the need for an organisational perspective about the incorporation of information technology, in a realistic way, with the available resources -materials, time and personnel. This depends on the global approaches of the school regarding their definition of what all the pupils at all school levels, and particularly in the final levels, need to know about information technology, both in knowledge and skills. It is also essential to analyse what this knowledge should be in each of the school subjects.

Secondly, the co-ordination of IT has a capital importance. By obtaining a sufficient operating level of the IT resources, which can be eased by the homogeneity of computer systems and programs, the co-ordinator ought to adopt a role of integrating and facilitating the participation of the staff, and of ensuring the decentralization of the use of technology. Its main roles should be support, simulation and diffusion of information. The IT co-ordinator should not assume the pedagogic responsibilities and obligations of the other teachers on the grounds of technological specialisation; every teacher ought to assume their own pedagogic responsibilities and not delegate them to the IT expert. The intervention of the IT co-ordinator in the classrooms ought to be limited only to when it is essential and, in exceptional cases, for particular supportive actions.

Finally, it is worth stressing the fact that integration of IT is both a school responsibility and an individual teacher responsibility. This means not only letting students work with IT and providing organisational measures to foster this work, but to rethink the contents and didactics of school subjects and the role of the teacher.

REFERENCES

1. Husen, T. (1988) *Research paradigms in education.* International Encyclopedia of Education Database, Pergamon Orbit InfoLine Ltd., London.

2. Martínez, J. B. (1990) *Hacia un enfoque interpretativo de la enseñanza.* Universidad de Granada.

3. Benedito, V. (1988) *La investigación didáctica.* In: Enciclopedia de Educación. Barcelona: Planeta.

4. Blomeyer, R. L.; Martin, C. D. (Eds.) (1991) *Case Studies in Computer Aided Learning.* The Falmer Press, London.

5. Cook, T. D.; Reichard, C. S. (1986) *Métodos cualitativos y cuantitativos en investigación evaluativa.* Morata, Madrid.

6. Elloit, J. (1989) *Pràctica, recerca y teoria en educació.* Eumo, Vic.

7. Gimeno, J., Pérez, A. (1983) *La enseñanza: su teoría y su práctica.* Akal, Madrid.

8. Taylor, S. J.; Bognan, R. (1992) *Introducción a los métodos cualitativos de investigación.* Paidós, Barcelona.

9. Guba, E. (1988) *Methodology of Naturalistic Inquiry in Educational Evaluation.* London: International Encyclopedia of Education Database, Pergamon Orbit InfoLine Ltd.

10. Goetz, J. P.; Lecompte, M. D. (1988) *Etnografía y diseño cualitativo en investigación educativa.* Madrid: Morata.

11. Stenhouse, L. (1988) *Case study methods.* International Encyclopedia of Education Databases, Pergamon Orbit InfoLine Ltd., London.

Jordi Quintana i Albalat is a graduate in pedagogy and primary teaching, a member of the Generalitat de Catalunya, and professor at the University of Barcelona. He has written several papers and publications on the didactics of mathematics and the educational integration of IT.

Ferran Ruiz i Tarragó is a physics graduate and secondary school teacher with postgraduate studies in teacher training. He directs the Informatics in Education Programme (PIE) of the Generalitat de Catalunya and is a member of IFIP Working Group 3.1. Ferran is the author of several papers and publications on the relationship between education and IT, including IFIP WG3.1 Guidelines of Good Practice: *Integration of Information Technology into Secondary Education: Main Issues and Perspectives.*

16

Factors affecting the use of computers in the classroom: four case studies

Wim Veen

Institute of Education
Utrecht University, The Netherlands

ABSTRACT

The decision of teachers whether or not to use computers depends on two basic categories of factors: factors at school level and factors at teacher level. However, teacher factors appear to be more significant than the factors at school level. Teachers have strong beliefs in respect to the content of their subject matter as well as to the pedagogy. The case-studies in one school (1989-1993) described here show that those beliefs appear to change only very slowly. Teachers adopt new media if they can use them in accordance with their existing beliefs and practice. The findings of this study coincide to a large extent with the results of many other studies of classroom practice.

Keywords: attitudes, case studies, classroom practice, innovation, pedagogy

INTRODUCTION

Between 1989 and 1991 four case-studies were undertaken to explore the day-to-day practice of teachers in one school who were introducing the use of computers in their classrooms. The teachers involved were provided with a computer at home, one computer and a transviewer in their classroom, and a fully fledged computer laboratory. The selected teachers included both average and experienced teachers who were novices in the field of information technology in education. The four teachers were given a standard in-service training course on computer-assisted teaching to bring them as a group to a same level of computer competence. After the course all further support by the researchers was withdrawn. The teachers were free to use or not to use the computers in their classrooms and at home. The main research question was

"If they have adequate resources, what are these teachers going to do with computers at school and at home and can their activities be explained?"

THE RESEARCH METHOD

To implement the case-study approach [1] different sources of data were used: questionnaires, teachers' diaries, different types of interviews, classroom observations and research group deliberations. (See Table 1.) The teachers kept semi-structured diaries; teachers were interviewed every two or three weeks about the entries in their diaries. Empirical data of classroom activities were collected by observations of computer assisted lessons. Finally, in-depth interviews were conducted every six months during which the teachers reflected on their latest activities.

	French	English	History	Geography	Informants	Total
Classroom observations	16	8	19	13	-	56
Diaries	182	62	167	118	-	529
Initial interviews	1	1	2	2	3	9
Interviews every fortnight	26	16	26	22	-	90
Interviews every six months	3	3	3	3	9	21
Follow-up interviews	2	2	2	2	6	14
Questionnaires	2	2	2	2	-	8
Plenary meetings in school	-	-	-	-	-	3

Table 1: Sources of data used in the research

For a better understanding of the context in which the teachers worked, the principal, the technical assistant and the IT co-ordinator were interviewed every six months. All interviews and classroom observations were audio-taped and transcribed. In order to eliminate bias in the collected data, all interview transcripts have been analysed by a different member of the research group. Researchers rotated classroom observations as well as interviews. In weekly meetings the research group discussed the collected data to reach a high degree of inter-subjectivity. In this way, from the fall of 1989 until January 1991 all the computer-related activities of the teachers were comprehensively described and analysed.

After this intensive period of data collection and analysis, there was a follow-up period of limited data collection from 1991 until June 1993; the teachers and the three informants were interviewed once a year.

The school at the start of the research

The four teachers (of French, English, Geography and History) all worked at a medium-sized secondary school (800 students) in an urbanized region in the centre of the Netherlands. The school enrols students of a variety of levels and was not seen as being unusual relative to other such general secondary schools. Certainly the school had no particular history of innovative practice. The school is known for its relatively traditional whole-group classroom teaching methods. Lessons are generally 45 minutes in length and students have three hours per week available for study help, homework supervision, and some free choice activities. The teaching staff is predominantly composed of teachers with a considerable amount of experience and mobility among the staff members is not usual.

The school had two computer rooms before the start of the research period. One was the standard configuration available in all secondary schools in The Netherlands, the other a small room with eight old microcomputers. Typically the computers were only used for the teaching of the required information technology course, consisting of general lessons in programming. The area of the school where the French teachers worked had an additional set of three computers, used for drill and practice exercises in French.

At the start of the research, the school had no definite policy about computer assisted learning. There was a teacher serving as a technical assistant for the use of the computers, but he had no formal task definition. However, by 1989 the new school principal held a positive view on computer-assisted learning. She believed that computers offered the possibility for more individualized learning and for more of an orientation toward inquiry learning within the school. Thus she supported the involvement of the school in the research, partly because she believed that the study could have a positive influence on the 'face of the school'. The teachers who participated in the study, however, had no release time available for the research.

The teachers at the start of the research

The four teachers all had good reputations. Three had about 20 years' teaching experience, while one, the history teacher, had 5 years' teaching experience. According to their own self-appraisals, roughly 80% of their lessons were "teacher-centred." Using a standard questionnaire consisting of 90 statements, the teachers could be compared to 95 Dutch teachers who had previously responded to the questionnaire on 11 issues concerning among others educational goals, beliefs, attitudes, willingness to innovate, job satisfaction and perception of appropriate teacher-student relationships. Based on the comparison with this norm group of teachers, the four teachers in this research appeared to be rather average teachers, although the English teacher scored very low on willingness to innovate. Certainly they did not belong among those who could be described as computer pioneers, and they did not have experience of computer assisted lessons. However, two of them used a computer at home for word processing.

Their views about the use of computers in teaching differed. The Geography and French teachers held positive views. They expected to use computers frequently for their teaching. The History teacher in contrast can be characterised as critical but positive. He had some critical comments about the way they should be used in education. The English teacher had several worries and fears about the use and implications of computers for his teaching situation.

THE RESULTS

After a year and a half of computer use the teachers had tried seven types of computer assisted lessons (CALs) as shown in Table 2 opposite. These types of lessons have been categorized according to the computer configurations and pedagogical settings involved. CAL organization types 1 to 7 relate to categories of computer assisted lesson delivery, including lesson types involving whole classroom teaching with one computer, (Type 1), different uses of two or three computers in the classroom by four or six students (Types 2 through 5), and use of the computer laboratory with whole or half classes (Types 6 and 7).

Types of CALs	French	English	Geography	History
CAL 1: The electronic blackboard	12	9	28	14
CAL 2: Computer use as working apart together	19	5	-	-
CAL 3: Rotating group work, with computer as one setting	5	-	-	-
CAL 4: All students using media, including computers	30	-	4	-
CAL 5: Voluntary extra work with computers	17	-	14	4
CAL 6: Computer lab work without support of technical assistant	23	2	13	15
CAL 7: Computer lab work with support of technical assistant	19	2	6	-
Totals	125	18	65	33

Table 2: Categorisation of the computer assisted lessons of the four teachers during the research period of August 1989 - January 1991

The teachers appeared to be quite different in their use of these different categories. Not only did the number of computer assisted lessons the teachers organised vary considerably, but the variety of types of computer assisted lessons also differed extensively among the teachers.

The French teacher
Of the four teachers, the French teacher used computers the most and he used them in all the different types of CALs. He said that he was conducting "experiments" in order to find out in what pedagogical settings he could best achieve the goals he had in mind. His goals were related to himself as a teacher and to his beliefs and skills with regard to the pedagogy of his subject matter. He used computers mainly for three purposes: to give information to the whole class, for remedial purposes and for drilling grammar and vocabulary. The CAL settings he used most frequently had pedagogical settings that he had mastered best in his regular teaching. Experience with these CALs gave him the opportunity to prepare himself for the use of CALs with more complex pedagogical settings, such as those involving group work (CAL Type 3). He had little experience with organizing group work; he thought it would require intensive changes to his teaching and be difficult to implement as he would have to manage different groups of students at the same time. As

far as his pedagogy was concerned, he considered himself as an advocate of a communicative approach of foreign language teaching. Nevertheless he also believed in drilling vocabulary and grammar. He frequently said:

"You know, repeating vocabulary and grammar is always useful to do. Computers can help with that."

In his attempts to integrate computers in his classroom practice, this experienced French teacher was, on the one hand, concerned with his own learning process, trying to enhance his pedagogical skills in a step by step manner. On the other hand, his belief in the usefulness of grammar and vocabulary strongly influenced his choice of instructional uses of computers. His general view on the educational value of computers played another part in his computer activities. He thought that computers motivated students and led them to work independently, which in his view should be the way for the future.

The English teacher
In contrast to the French teacher, the English teacher used computers only sporadically. He started by using the computer several times in whole-group classroom teaching situations "for fun" by doing word-gaming. But technical limitations of the LCD-screen stopped him from further use of the computer for whole-group classroom teaching. Occasionally he later used his classroom computer for remedial purposes. Students who had poor scores on some tests could work two by two on the computer while the rest of the students worked in a ordinary whole-group classroom setting. These computer activities, he said, were very disappointing. His classroom order was disturbed considerably by the fact that he had the computer using students rotate approximately every ten minutes. Students were excited and talked a lot.

The teacher initially thought that computers could be useful for writing. In his view it was compulsory however, to train students in the technical skills needed for word processing first. He thus started a word processing course in the computer lab, but here again his experience was negative. By the end of 1989 he stopped his teaching activities with computers in school. He attributed his negative experience to a lack of motivation on the part of the students and to the size of his classes. But most of all, he complained that the software he was familiar with did not suit the textbook he was using closely enough. Moreover, in his opinion, the software did not add any educational value to his lessons. Finally, using computers did not suit his "teacher-centred" teaching style. He considered computers only to be useful in very small classes with very motivated students.

This teacher clearly shows the predominant influence of beliefs and skills in the subsequent uses a teacher makes of computers. He had little flexibility in communicative and pedagogical skills and notably he had fears of loosing control of students' activities. Moreover, he was convinced that his "teacher led" lessons were more effective than computer assisted lessons. Using computers made him deviate too much from his existing teaching skills and he did not succeed in finding a way to integrate computers in his existing teaching style. As a consequence he decided to give up trying to use computers for teaching.

The Geography teacher

The Geography teacher made use of his classroom computer and transviewer frequently. During whole-group classroom teaching he used some simulations but most of all he used the computer in his classroom as an electronic blackboard. After getting acquainted with a geographical database on national planning, he very frequently supported his instructional talk by showing graphic displays of data (maps, graphs). In that way, he said, he fitted the computer seamlessly into his expository teaching style.

He also regularly went to the computer lab with his classes to work on geographically oriented computer simulations. As these lessons represented new pedagogical settings for the teacher, he prepared them thoroughly. He wanted to keep control over the students' learning process. Students received precise instructions on every step they had to make as "he did not want them to get lost in the courseware." He was surprised to find out that students were motivated and quickly found their own way in the software.

Over time his preparation time of computer assisted lessons diminished considerably and he started using the accompanying written materials provided by the publisher. When students asked the teacher for courseware to help them to prepare for their examinations, he enabled them to use the classroom computer or the computer lab during recreation time or after school time.

The Geography teacher is another example of fitting computers into existing teaching skills. He was most pleased when using computers in his teacher centred lessons. Those were the lessons he which he was very familiar, so he felt confident in this pedagogical setting. However, he consciously enhanced his teaching experience by working in the computer lab. He said that he was aware of his predominant use of whole-group classroom teaching and he wanted to reduce its frequency.

The History teacher

Although the History teacher was critical about the use of computers in teaching, he admitted he saw some advantages for using computers in his lessons. He said that computers stimulate an explorative learning approach and motivate students in history. Nevertheless he used the computer less as part of his regular routine than the French and Geography teachers. In addition to what he saw as a serious lack of software in his domain and his concern about fitting courseware in the curriculum, his relatively little use of computers can be explained by his preference for expository teaching. He held a strong pedagogical view that a history teacher should always start the learning process of students by telling them the general social-economic and political backgrounds to historical events:

> "I don't believe that students will get a real understanding of history by analyzing piles of original documents or texts and reading simple facts. For the historical insight in events students need first a general background of the historical epoch in which events occur."

And this general background was, in his opinion, lacking in historical course-ware. Apparently, the educational paradigm of the History teacher did not match the paradigm of the courseware available to him, courseware that was oriented toward inquiry learning. On the one hand, the History teacher fitted the use of courseware into his pedagogical view on his subject matter. On the other hand, however, he also tried to enhance his teaching skills by working in the computer lab.

The home use of computers

All the teachers used their computers at home frequently and intensively for word processing. The four teachers considered this computer application as very useful and time-saving. This use appealed to their need for efficiency. The preparation of lessons took the Geography and French teacher about one-fifth of the time spent at home. The Geography teacher prepared his lessons carefully, especially in the beginning. On the contrary, the History teacher preferred not to spend too long in preparing his lessons and liked to try out things and see the results in practice.

Although the teachers showed significant differences in computer uses in the classroom, they all spent about two-and-a-half hours a week at their computers at home. The four teachers felt this to be an acceptable time in their normal weekly work load.

Two years after

As far as the use of computers for teaching is concerned the teachers significantly reduced their activities between 1991 and 1993. One of the reasons for this decline in activities was that during these years the researchers withdrew from school and data were collected only by interviews once a year. The use of computers in teaching situations that occurred were predominantly related to the use of drill and practice software such as the training of grammar, vocabulary and preparation for final exams. According to their preferred teaching styles, the teachers used computers in teaching settings with which they were already familiar. Thus, the History teacher let his students use software to help them prepare their exams, and the French teacher continued using computers in his classroom with four to six computer-using students. The Geography teacher met serious problems in getting new hardware for his expository teaching which resulted in no use of the computer in his classroom.

Although the English teacher bought a new textbook in which the use of software was an integrated teaching activity, he found a way to avoid computer assisted teaching situations in which he feared to loose control over his students. He gave the software to his students who were to work with it at home.

After four years of computer experience, the teachers' views on computer assisted teaching appeared hardly to have changed. Their beliefs on what should be in the curriculum of their subject matter and the way this content should be taught determined to a large extent their uses of computers. Although the teachers said they wanted to reduce their central role in their lessons, none of them had succeeded in putting this into practice by using the computer.

FACTORS INFLUENCING THE TEACHERS' USES OF COMPUTERS

After analyses of the case studies, two major categories of factors explaining the uptake of computers by the four teachers have been discerned: factors at the school level and factors at the teacher level.

School-level factors

School factors played an important role in how the teachers made use of their computers. First, the support of the technical assistant was essential for the teachers. Not only did they count on his support for tasks such as making copies of software, they more significantly relied on his collaboration during many of their computer assisted lessons in the computer lab. Secondly, the principal played an important role. She held a positive view on information technology in education and she provided the necessary technical support for the teachers by allocating a technical assistant for twenty hours a week. Moreover, she established an IT committee to be a platform for discussion and computer related policy making. Financial support was given to the teachers so that they could purchase the required software. And last but not least, the principal gave moral support through informal talks with the teachers showing her commitment and interest in their efforts.

Teacher-level factors
However, teacher factors outweighed the school factors in explaining how the teachers' used computers. These teacher level factors could be grouped into two sub-categories: beliefs and skills. Most important of these were teachers' beliefs regarding what should be in the curricula (content) and the way in which the subject should be taught (pedagogy).

Teachers also had beliefs regarding their role in the classroom and the corresponding classroom activities. Finally they had personal views on education and on themselves as teachers that also influenced what they did with the computers.

The skills of teachers that most influenced their uses of computers were those skills related to their competence in managing classroom activities, to their pedagogical skills and, less importantly, to their computer handling technical skills. These inter-related beliefs and skills can briefly be described of as "routines".

In summary, the uptake of computers in the classroom was strongly affected by the beliefs and skills or 'routines' of the four teachers. They tried to fit information technology into pedagogical approaches consistent with their beliefs and their skills. Although three of the teachers tried to enhance their skills, their beliefs were hardly changed by the influence of information technology.

RESULTS OF OTHER RESEARCH

Comparing the results of our study with that of other research, we have found a considerable consistency in findings. Some common findings can be identified.

Persistence of beliefs
Ten years after Fullan [2] stated that "educational change depends on what teachers think and do" (p. 107), he stressed the importance of this

statement by identifying the beliefs of teachers as very critical elements in educational innovations. Teachers hold views that persist during innovations. Educational change is a slow process and teachers need time to gain experience with computers. In a nation-wide survey simultaneously with a number of case studies, Watson [3] concluded that teachers' beliefs and their resulting teaching styles changed very little over time:

> "... teachers using IT often considered that computers were to be used to complement rather than change existing pedagogic practice, whether it be 'traditional' or 'progressive' (p. 160)".

In the majority of their use of computers, the teachers in our case studies did little that could be described as exploiting the overwhelming educational possibilities of information technology so often described in literature [4, 5 and 6]. They did not even consider these possibilities. Instead, they started using the computers for their own convenience by doing word processing. Then they carefully set out for the use of computers in classrooms where they varied pedagogical conditions as less as possible. Gradually, they experimented with pedagogical settings that were more difficult for them.

The persistence of beliefs of teachers during innovations is also supported by Olson & Eaton [7], Martin [8], Carmichael et al. [9] and Rubin & Bruce [10]. These authors all stress the influence of existing classroom routines. Teachers have standardized behaviours and it is "through such routines that teachers express their ideas of how classroom life is to be conducted" (Olson [11]).

Fitting into beliefs and skills

As a consequence of the predominant influence of teachers' beliefs on educational change, many authors support our view that innovative activities should fit existing beliefs and skills of teachers. As Olson [11] states:

> "...We can now see that classroom routines are not what computers will replace, they are where computers must fit if they are to be useful to teachers..."

Wiske et al. [12, 13 and 14] found that beliefs related to subject specific pedagogy are critical in innovative activities of teachers. In their view, teachers' need to adapt innovations stems from their preferred teaching styles and values. This need for adaptation is also reported by Rubin & Bruce [10] who conclude that in their study teachers "create new practices that reflect complex and situation-specific combinations of old and new approaches" (p. 2).

This orientation might upset some advocates of information technology. However, only by acquiring positive experiences with computers in a way that teachers do appreciate, is there a chance that their newly developed skills will start to change their beliefs. For any educational innovator it is important to realise that it is not the view of the innovator about the merits of the innovation that matters, but rather it is the view of the teachers about the innovation that is critical. If teachers start using computers for drill and practice only, it is probably that use that fits their routines best. Their learning process should not be disturbed by telling them that doing drill and practice with computers is only a poor application of information technology. Perhaps it will be only after two or three years that teachers can gradually enhance their routines and handle more complex applications of information technology.

Importance of the support of the principal
A third conclusion that emerged from our case studies is also well supported in the literature: the implementation of innovations should not be the task of individual teachers only [15, 16, 17 and 18]. The school management also has to be involved financially as well as through policy and moral support. As change is a slow process, it is essential that school managers look far ahead and persist in their support of change. In our research, two years of innovative activities appeared only to be the beginning for an integrated uptake of computers for two of the teachers. School managers should realise that not all teachers in the school are equally interested in educational change, particularly change involving information technology. A differentiated bottom-up approach for the different subject-area departments in the school seems to fit the school reality the best.

Importance of school culture
By the end of the eighties, a critical issue in literature about educational change has been the influence of the school culture on innovations [7, 13, 15, 17, 19, 20]. Sheingold et. al. [15] observed that in schools with a "do your own thing ethic", innovative activities were likely to stay restricted to some individual pioneers in the school. Schools with a culture of collaboration between teachers gave way to the pioneers to share their ideas with other teachers. The results of our case studies do support this findings in that the teachers worked within a non-collaborative school

culture in which colleagues did not give any feedback on their enthusiastic efforts to implement computers in their instructional practice.

A DESCRIPTIVE MODEL

Factors affecting the uptake of computers in the classroom appear to be strongly interrelated. In order to visualize the relations between the different factors, we have developed a descriptive model (see below).

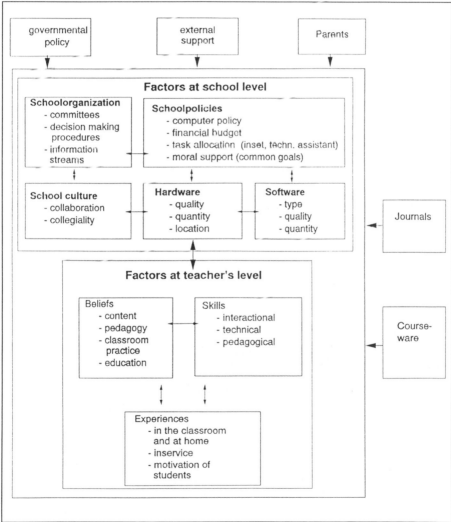

Figure 1: A descriptive model of factors affecting computer use in schools.

The model is based on our case studies as well as the synthesis of 32 other case studies [21, 22 and 23]. It gives a comprehensive overview of all the factors that were considered by the researchers as relevant and about which exists a certain consensus. Although the model gives a clear picture of the interrelationship of factors it is limited by the fact that no weight of factors can be discerned. Nevertheless, we should like to propose as a hypothesis that the model could offer a framework for analyzing the chances of success of the implementation of computers in schools.

REFERENCES

1. Yin, R. K. (1989) *Case Study Research, Design and Methods.* Newbury Park, CA: Sage Publications.

2. Fullan, M.G. (1982) *The meaning of educational change.* Toronto: OISE Press, and New York: Teachers College Press.

3. Watson, D.M., ed. (1993). *The Impact Report: An evaluation of the impact of information technology on children's achievements in primary and secondary schools.* London: Department for Education and King's College London, Centre for Educational Studies.

4. Summers, E.G. (1985) *Microcomputers as a New Technological Innovation in Education: Growth of the Related Journal Literature.* Educational Technology, August 1985, 5-14.

5. Whiting, J., (1985) *New Directions in Educational Computing: Coming Changes in Software and Teaching Strategies to Optimize Learning.* Educational Technology, September 1985, pp. 18-21.

6. Norton, P. (1985) *An Agenda for Technology and Education: Eight Imperatives.* Educational Technology, January 1985, pp. 15-20.

7. Olson, J., Eaton, S. (1986) *Case Studies of Microcomputers in the Classroom, Questions for Curriculum and Teacher Education.* Toronto: OISE Publication Services.

8. Martin, C.D. (1991) *Stakeholders/perspectives on the implementation of micros in a schooldistrict.* In: R.L. Blomeyer Jr. & C.D Martin (Eds.), Case studies in computer aided learning, pp. 169-221. London: The Falmer Press.

9. Carmichael, H.W., Burnett, J.D., Higginson, W.C., Moore, B.G. & Pollard, P.J. (1985) *Computers, children and classrooms: A multisite evaluation of the creative use of micro-computers by elementary school children.* Ontario, TO: Ministry of Education.

10. Rubin, A., & Bruce, B. (1990) *Alternate realizaitions of purpose in computer-supported writing (technical report No. 492).* Champaign: University of Illinois, Center for the study of reading (ERIC Document Reproduction Service No. ED 314 760).

11. Olson, J. (1992) *Trojan horse or teacher's pet? Computers and the teacher's influence.* International Journal of Educational Research, 17 (1), 77-85.

12. Wiske, M.S. et al. (1988) *How technology affects teaching.* Cambridge, MA: Harvard Graduate School of Education and Education Development Center. Educational Technology Center.

13. Shepard, J.W., Wiske, M.S. (1989) *Extending technological innovations in schools: Three case studies and analysis.* Cambridge, MA: Harvard School of Education, Educational Technology Center (ERIC Document Reproduction Service No. ED 303 372).

14. Wiske, M.S., Houde, R. (1988) *From recitation to construction: Teachers change with new technologies.* Cambridge, MA: Harvard Graduate School of Education, Educational Technology Center.

15. Sheingold, K., Kane, J.H., & Endreweit, M.E. (1983) *Microcomputer use in schools: Developing a research agenda.* Harvard Educational Review, 53 (4), 412-432.

16. Cox, M., Rhodes, V., & Hall, J. (1988) *The use of computer assisted learning in primary schools: Some factors affecting the uptake.* Computers and Education, 12 (1), 173-178.

17. Fullan, M.G. (1992a) *Successful School Improvement.* Buckingham: Open University Press.

18. Fullan, M.G. (1992b) *Teacher Development and Educational Change.* London: The Falmer Press.

19. Sarason, S.B. (1982) *The culture of the school and the problem of change* (2nd ed.). Boston: Allyn and Bacon.

20. Sikes, P.J. (1992) *Imposed change and the experienced teacher.* In: M. Fullan & A. Hargereaves (Eds.), Teacher Development and educational change, pp. 36-55. London: The Falmer Press.

21. Veen, W. (1993) *How teachers use computers in instructional practice: Four case studies in a Dutch secondary school.* In: Moonen, J.J. & Collis, B (eds.), Computers and Education, **21** (1/2), 1-8.

22. Veen, W., Vogelzang, F., Neut v.d. A. C., Spoon, P.P. (1992) *How computers are used in the instructional practice.* In: Plomp, T., Pieters, J. & Feteris, A. (eds.), Proceedings of the European Conference on Educational Research (ECER) **2,** 446-454. Enschede: Twente University Press.

23. Veen, W. (1994) *Computerondersteunde docenten; de rol van de docent bij de invoering van computers in de klaspraktijk. (Computer assisted Teachers; The Teacher's role in the implementation of computers in the instructional practice).* Diss. De Lier: Academisch Boeken Centrum.

Wim Veen has been working at the Institute of Education at the University of Utrecht since 1978. For ten years he has been a trainer of geography teachers in both initial and in-service training. In 1986 he created a research group on information technology. He has been involved in a number of IT projects related to the implementation of computers in schools, the development and translation of coursewear and the use of telecommunications. Wim is currently involved in the research on policy and planning issues with respect to the support of telecommunications in education.

17

Integrating IT into teaching - struggling with windmills

Lilliam Hurst

Centre Informatique Pédagogique
Genève, Switzerland

ABSTRACT

As soon as I started learning how to use the New Information Technology, I started trying to figure out ways to allow my students to enter the same marvellous world I was exploring. The windmills I had to struggle with were varied: the headmaster had to be convinced to fund the project ; my technology-resistant students had to be urged into the computer-room, then urged out again when they got hooked in turn; the colleagues had to be won over - this was a particularly resistant windmill. Much time was spent learning how to use the technology, acquiring new software, preparing menus to make the students' life easier, but that wasn't really a struggle. Other windmills reared up: the technical experts had to be convinced that having a language teacher in the "holy of holies" was not only a viable, but necessary; the librarian, who could have been another difficult windmill, turned out to be an ally. It was she who insisted that the computer should be in the library, as one tool among others. But windmills, if properly handled, can turn out to be allies in the end; and for the time being, they are providing me with energy.

Keywords: teaching methods, curriculum, communication, integration, motivation

INTRODUCTION

"If I had the students 12 hours a week, I would try to integrate IT into my teaching" said one colleague. Her husband commented further: "My wife will not learn to use a computer because it's senseless to use something which is so complex - why, she would only ever be able to master about 5% of the entire process!" I haven't precisely observed her avoiding the use of her brain, however, and I have been told that we only use, at any given time, a tiny percentage of its capabilities...

I am a language teacher and my main professional purpose is and must remain language learning. I work at three sites in the State system of Canton Geneva - Collège Claparède, Collège pour Adultes and the Centre Informatique Pédagogique. In this paper I will address some of the issues I faced when starting to use computers for language learning. By tackling these issues, a wide range of curriculum related activities are now active on these sites.

ENCOURAGING TEACHERS TO USE IT

One of my aims as a teacher has been to become as contagious as possible, to persuade more and more of my colleagues to become involved in work with new information technology.

To do this, I have used the pebble-in-the pond approach. I started using Computer Assisted Language Learning (CALL) with my own students almost as soon as I had had been shown which button switched the machine on. My students were given instructions to enable them to use the programs I was installing on the hard disk of the computer in the library as fast as I could buy them. At first we used programs written by others, including exercises and feedback which were appropriate for our needs. Being eager to improve their English, however, my students swaggered into the library with their instruction-sheets, and at once became the envy of others, who borrowed the sheets, and started doing the work. This led to their asking their teachers why they weren't doing the same thing; teachers soon came to ask me to show them how. So the pebble made one circle of ripples - at least at my school.

Not everything went according to plan. The first windmill I had to struggle against was teacher indifference, not to say hostility. A first contact could be followed up by a suggestion that I be invited to a session with half-groups to show them what software was available, and how it worked. In many cases, the students were the motivators, which is an interesting reversal of the common situation.

The headmaster had to be convinced that, contrary to his original expectations, the extra time I was spending doing something which he hadn't initiated, had to be paid for. In a time of increasing budget restrictions, he had to make and keep a place in his budget for something which hadn't, until then, been high on his list of priorities and the utility of which he could only vaguely see. But, he worked on the principle: "If you can't lead, follow; and if you can't follow, then get out of the way".

At the beginning of each school year I have to convince my headmaster that he cannot afford not to have the new technologies in his budget for foreign language instruction. It takes a continuing effort to keep this weekly non-teaching hour in the budget, and in my contract. In return, I organise all the software for English, German and Italian, for my own use and for my language colleagues, who rely on this help. I would say that this is one of the hardest windmills that I have had to struggle with.

Students' response

Students have proved to be the most rewarding windmill to struggle with. Their responses have ranged from the unflattering: "I think it makes a change in the rhythm and it is a good way to teach us things that are boring to learn. It is also a chance for us to learn how a computer works. It's very useful" or "I think it's very good because we can work all the time (reference to the self-access system at Claparède), and it's very good for the future, because, actually, computers are very important in the world"; to an "allergic" student who declared that computers were not his "cup of tea", and that he wanted nothing to do with them.

I responded by continuing to give homework which was available only on the computer, and on which there was a test at the end of a certain period. This approach was more than vindicated by a visit from this 'allergic' student, during his second year with me. He wanted to know why I hadn't continued creating exercises for the fourth-year pupils - he had finished all of them, and it wasn't very pleasant to have to do exercises twice.

Key relationships

Existing practice
In our schools in Geneva, there is a technically oriented teacher, usually a teacher of computer science who is responsible for the upkeep and

maintenance of the Local Area Network (LAN).   That person can effectively keep other teachers, including language teachers, out of the computer lab, and a good relationship is essential with them if the projects are to succeed.

Relations with successive RAs
The first Responsable d'Atelier (RA) with whom I had to deal was Philippe Drompt, who was not only a computer genius, but was also convinced of the importance of language teachers for the future of the integration of the new technologies into the entire school curriculum. He provided key mentorship and help while I was myself learning about IT use.   The current RA at our school is Derek White, who is unfailingly patient with requests for help. We feel that he is "on our side" and that is half the battle.   Among other things, the computer department in my school finds it more convenient to have me attend all their meetings, so as to avoid having to transmit messages about what took place.

Relations with the librarian
The librarian was a key person for the installation of a stand-alone computer in the library, so that the students could see that it was simply one tool among many. Sadly, however, the stand-alone Mac was stolen from the library, and so removed from the students' reach.

Relations with colleagues
It seemed natural to share what I was learning with my colleagues. If it was making my life simpler, there was every reason in the world to allow them to have their lives simplified too.

When colleagues ask me how to use the new technologies, I first show them how the programmes work. Then, I hand out timetable sheets with my available hours so that they can issue official invitations to their half-group sessions. This means that when they have an hour slotted for smaller group-work, I go along and show the students what is available on the computer. It also allows me to tell the students that it is my work they will be destroying if they sabotage the programs, and to request their assistance in finding any mistakes I have made. It gives them a sense of responsibility and power to realize that they are being asked to work with me.

Colleagues from other schools have also asked me to show them our material. We usually have an after-school session at my school, so that they can explore the programmes; then I give them the menu-system I wrote so that they can start by doing the same thing. The data portion of our software, that is the exercises written by us, are of course available freely for others to use, with the request that they let us know if something can be improved upon. This has proven to be a problem at times, when colleagues call up at all hours and, without asking if the time is convenient, pitch right in with their comments, sometimes leaving me in the dark as to what they are talking about.

Curriculum activities

Below are some examples of a variety of curriculum examples that have been started, I have used these "success stories" in turn to attempt to make further ripples by providing examples of successful activities for colleagues to explore.

These activities involve use of the computer, but need not be exclusively limited to CALL. Many of these success stories have been collected and a monograph will be available by the time this paper goes to press.

Short story writing: using an activity taken from Ron White's *Writing* in Oxford Supplementary Skills [1], we have held simulation and incremental computer write-up sessions, leading to the completion of a 1000-word short story by the students.

Made-to-order exercises: we are constantly creating exercises that mesh with what we do in class, so that the students can reinforce and follow up work in class with sessions at the computer.

First Certificate request from pupils: the students who wish to do the First Certificate of English examination with the University of Cambridge Syndicate can do exercises taken from past exams, or prepare for the exam by using one of the few acceptable dedicated programmes in the field.

Diskette loans to adult students at Collège pour Adultes: until there is network access for all, I have instituted a disk lending system with my students at the Collège. They begin the year with student programs on disk-exercises relating to the first part of the curriculum; they return the disks to me around the middle of the year, and carry on with the second half of the curriculum also on disk. This is burdensome for me, as it involves a great deal of collating and copying, but it's better than nothing.

WDS disks to colleagues: my colleagues at the Collège pour Adultes have requested the vocabulary lists that we have typed in for the text books in use there. The student version of the program used for this has been shared with the colleagues requesting this. They in turn have found students who are computer literate enough to print out vocabulary lists for the entire class.

BBS "agony aunt" services: with one 3rd-year classics class, we have a local Bulletin Board Network group known as BENHURST. The students who belong to BENHURST have to send their Agony Aunt a message every ten days, with a query related to something we have been doing in class, or to a question that is still unclarified regarding English grammar, or, on any subject, as long as the question is in English. I then correct their English and answer their question. It is simply a modern version of the old classic Journal, whereby notebooks were filled with an on-going correspondence between teacher and learner. I have simply adapted this technique to the new technology. The success of this exchange has kept my mailbox filled to overflowing. The use of BENHURST has made available a cohort of sensitised students for work in 1992 described below.

A communications network
'Learning Circles' is a communication network consisting of 7 to 10 teams from different schools, constituted according to age, level, and interests of the pupils. On joining, the team can choose the area of the curriculum it wishes to work with. My school's choice was from:

- Journalism & Language Arts (Computer Chronicles)
- Creative & Expository Writing (Mind Works)
- History & Geography (Places and Perspectives)
- Social Studies & Current Issues (Global Issues)
- Social Studies (Society's Problems)
- Science & Current Issues (Energy/Environment)

Messages are sent automatically to a central AT&T computer by dedicated software which then redistributes them to the electronic 'mailbox' at each participating site. An interactive 'Learning Kit' helps the teacher to cope with the software, and on-line help is only a reverse-charge phone-call away. Within each circle, a volunteer Circle Co-ordinator serves as a role model, provides curriculum guidance and monitors Learning Circle progress.

Each class also chooses a 'project' to be more closely associated with, for example, sending out questionnaires with subsequent analysis or doing local research, with a report on the findings sent out to other members of the Circle.   At the end of the session, each learning circle publishes a 'Circle Publication', under the aegis of the Co-ordinator.

Alongside the 'serious' work pertaining to their project, the students share personal experiences with their peers; this is one of their favourite times on the network, and several participating students carry on corresponding via 'real-mail' after the real session ends.

Social and political influences

In 1992 a group of 17/18 year-old students wanted in some way to make their mark in their network circle. Because there had been a scandal in Geneva, when the Council of State (the Chief Executive Body) had expelled a group of Albanians who were seeking political asylum, the students wanted to be able to discuss the event with some of the people involved in the decision. They therefore chose the theme Illegal Immigration.

Having chosen their theme, they composed a questionnaire - in French which then had to be translated into English. This was sent to our partners at the eight other schools of our Learning Circle, with the request that the questions be asked of anyone in power . Since most of the partner-schools were in areas far from the problems of illegal immigration, we did not receive any responses from them on this, but that was not viewed as a problem.

I wrote letters, on behalf of the students, to one right-wing politician in the Grand Conseil, the Legislative Branch of government, one moderate socialist, and, to round the picture out, to the President of the Department of Justice and Police, which had been the one most actively concerned by the expulsions. Those contacted all agreed to meet with the students, with the President of the J & P Department sending them his Secretary General, due to lack of time.

In the event, the right-wing member of parliament, thrilled with what he saw as the opportunity to air his views, even offered to come to my class to explain to the students why the expulsion was not only necessary, but good, and in general why his party was the one they should adhere to. For various ethical reasons, we refused, but it was good for the students to realize that there was information to be garnered from the world outside school, and that they could not expect everything to automatically come to them.

Obtaining and translating the answers for our partners in the Learning Circle was a truly rewarding part of the activity. The students worked outside class time to do this and were very concerned that their written work should be checked so that it was a true reflection of what they had been told. I was being transformed into a facilitator - which I much preferred to the role of judge for the Learning Circle.

Once the questions had been asked (in French) and the answers faithfully jotted down, the students returned, and one student from each team of answer-gatherers became the head of another team of translators. The questions and answers had to be distributed to our partners in the Learning Circle.

CONCLUSION

The pretext for doing all of this is the vehicular language, English. The main reason for our participating in any activity must, in my view, be subordinated to what is supposed to be my main goal - making my students independent and fluent in their use of English. It is fortunate that so many activities are available in this target language. The fact that most of the students in my school (and, I rather suspect, in many schools in non-anglophone countries) are eager to acquire competence in English simply makes the entire effort more feasible.

REFERENCE

1. White, R. (1987) *Writing:* Oxford Supplementary Skills, Oxford University Press.

Lilliam Hurst moved to Switzerland in 1966 on a 'temporary' basis after obtaining her B.A. from a College in California and is still there. She obtained a Licence ès Lettres from the University of Geneva in 1972, and has been teaching since 1976 (having obtained her Certification along the way). A British Council Specialist Course on CALL was the starting point for her interest in computer assisted learning and teaching, and her students' needs have drawn her ever further into the fascinating fields of self-access centres, telecomputing, and class-networking.

18

Effects of learners' characteristics and instructional guidance on computer assisted learning

Qi Chen

Department of Psychology
Beijing Normal University, China

ABSTRACT

Research has shown that only under certain conditions can CAI or other media function more effectively. This paper summarizes, on the basis of our several studies in secondary schools, how the variables such as instructional guidance, learner's learning style, and the extent of learner control moderate the effects of computer use in classrooms. Learners are the masters in the learning process and the inner factor of their mental development. Only through the learners' mediation can the technical potential be realized. Only those teachers who have mastered an understanding of the students' learning processes are aware of the conditions under which the new technology can play its role efficiently.

Keywords: teaching methods, courseware, learning models, learning skill, motivation.

INTRODUCTION

The Chinese education system

In the Chinese system, the central government outlines educational principle and policies through the State Education Commission which is the highest educational administrative institution and is controlled directly by the State Council. After many years of exploration, and especially the recent decade's experience, the Chinese education system has been undergoing a process of continuous reform and readjustment. A whole new set of systems has come into being. General education is mainly publicly funded; however, privately-funded schools have been permitted.

Although the pre-school education has been developed rapidly in China, it is not compulsory. The nine-year compulsory education that includes primary and junior secondary has been implemented step by step in a planned way, starting from 1985 [1]. The schooling system, including primary, junior and senior secondary, is mainly a 6-3-3-year system. In some regions, 5-4-3 or 5-3-3-system is implemented instead. The policy or principle for education is to enable students to get all-round development in moral, intellectual, physical and aesthetical aspects. The graduates of junior high schools must take an entrance examination held by local educational bureaus to enter senior secondary education, including professional and polytechnic schools. The graduates of senior secondary have to take the unified national admissions examination for high education. Because of a limited quota for studying, only a small percentage of senior high graduates are admitted to the universities and colleges.

Curriculum reform has been progressing rapidly. It used to be unified for the whole country either in syllabus or in textbook. Now it has been changed to "one syllabus but multitexts". One syllabus means the basic requirement for each subject should be met to ensure the quality of schooling, while multitexts meet different needs of different areas. Since China is such a big country and the developments in different areas have been so unbalanced, this reform policy gives local educational bureaus more flexibility to design their own texts to meet the local needs. The curriculum structure in senior secondary education also has been reformed to be more flexible. More free elective courses which are more practical and related to vocational education have been offered instead of some required courses. In addition, some activity courses have been offered to meet the needs of situated learning.

Description of educational computing in the classroom

In China, educational computing in secondary schools dates back to 1982. Learning about the computer, mainly learning programming languages, has been the core of computer education in secondary schools [2]. Since 1986, the State Education Commission along with other related institutions, has realized that to integrate computer into curricula should be the trend of educational computing [3]. While financial funds were being allocated for software development, special teams were organized to develop educational software. At the same time an educational software evaluation committee was established. As a result, computer assisted instruction was greatly encouraged. During its rapid development, educators also realized that as the core of modern technology, the use of computer in instruction would be an inevitable trend.

Nevertheless, there have been several different perspectives on the functions of CAI in schools. Schools which had advanced educational perceptions and ideas about the educational reform encouraged their best teachers to use the computer as an instructional medium to improve their teaching quality. The best teachers, programmers, and curriculum designers were brought together to develop courseware. Pilot studies and experiments were conducted, and, after those pilot studies, experimentation had been conducted, and, after these pilot studies, further experimentation took place and this led to an expansion of the implementation. Unfortunately, only a few schools held this type of perspective on CAI.

Some schools or regions are very enthusiastic towards CAI and adopted it completely. But they regarded information technology as a "fashion", they had no theoretical concepts about how educational technology should function well and what has happened in the world of information technology. The effects of CAI have often been so exaggerated as a kind of panacea for educational reform that CAI itself has become not an approach but an objective in teaching. It seems they believed that as long as CAI was used, its potential power would be automatically harnessed. Enthusiasm towards computer use in education should be protected and encouraged, but it is possible that some theoretical training about CAI should be given to those educators who hold this unquestioning enthusiasm.

A third perspective is found in those schools that are well-equipped with computers but without any knowledge or enthusiasm towards educational computing at all. Sometimes they just exhibit their computers to the visitors, or only teach about the computer.

RESEARCH FINDINGS

The impact of integrating information technology into education has been investigated in the past three decades. The initial positive fever for computer education has now been replaced by a more tempered attitude. It appears that CAI is relatively more effective only under certain conditions. Variables pertaining to learners, instruction, and software should be examined systematically [4]. This paper summarizes from our several studies, how those variables, such as learning style, the extent of learners' control and instructional guidance, could moderate the effects of educational computing.

Teacher's guidance and learner's characteristics

To examine how the extent of teacher's guidance moderates the effects of CAI, a series of studies on Computer Assisted English Learning were conducted in three secondary schools [5]; GuangDong Experimental Middle School in GuangZhou, GuangDong, P.R.C., (hereafter referred to as School A), ZhuHai # 1 Middle School in ZhuHai, GuangDong, P.R.C. (referred to as School B), and Workers' Children School in Macau (referred to as school C). Two Form Four classes, one for experimental group and one for control, were randomly selected in each of these schools as the study target. The experimental group used the computer assisted English learning, while the control group was taught the same content by an English teacher. The same educational software was used in each of these schools' experimental groups, but with different degrees of instructional guidance. In School A, the teacher gave the meaning of most of the vocabularies to the students (Most Guidance), while in school C, the teacher was told to let the students themselves explore the meaning of the vocabularies (Least Guidance). In School B, the instructor provided the students with the meaning of the key words only (Moderate Guidance). The first stage results showed that the experimental group in School C exhibited significant differences from their control counterparts in their final achievement. There were no significant differences in their achievement between the experimental and control classes in the other two schools.

Examination of the experimental procedure indicated that, despite the initial effort to standardize the conditions and procedure of the study in these three schools, students in School A appeared to be more motivated and of higher standard to start with, and that the experimenter (i.e., teacher) in School B was less enthusiastic than his counterparts in the other two schools. Consequently two supplementary studies were conducted. In School A, software of a more difficult level than the original one was used. In School B, the experiment was conducted by an experimenter whose instructional behaviour was trained to be comparable to that of the other two schools. The final achievement showed some improvement of those two experimental groups, though the differences were still not significant. Our preliminary conclusion for this aspect of teachers is that less guidance could induce greater student achievement [5].

Contrary to what was found for achievement, a measure of attitude indicated, that the target students in School A and B harboured a much more positive attitude and expectation toward Computer Assisted Learning than those in School C, though their achievements lagged far behind. They believed that the more they learned, the more their capacity for self-learning would be fostered , and the more they were willing to undergo Computer Assisted Learning in future. We found that the characteristics of students, including cultural background and level of motivation, are as relevant when considering the use of CAI as the extent of teacher guidance.

The students in School A and B were strictly selected by a unified entrance examination. Both students in experimental group and control group were well-motivated and had quite a high level in English study. Because of the high starting point of those students, it might be hard to show the differences in English achievement using different means in merely six weeks. More needs to be known about the processes leading to our research result that more guidance induced less visible achievement but a more positive attitude. Perhaps with more guidance, students felt the task was more attainable, resulting in a more assured confidence in learning with computer. It is also plausible that with more guidance, students could focus on other aspects of learning with the computer, thus raising their interest in such medium of learning.

This study focussed on one aspect of instruction. The extent that other aspects of instruction, such as the time given to students to interact with the computer, which might affect students' learning, attitude, and perception in different ways also should be examined. Equally important are variables related to the learners, which may interact with style of instruction to affect their learning and attitude.

The extent of learner control and cognitive style

To examine the effects of types of instructional control and cognitive style on tutorial CAI, a three by three matrix design was used; program adaptive control/ learner control/ learner with advice control were considered against field dependence/ middle level/ field independence [6]. Results indicated that learner control and learner with advice control were more effective than program adaptive control. Although there were no significant differences between these two effective groups, learner with advice control was slightly better than learner control.

Learning with the computer, as a relatively newer medium of instruction, appears to call for more participation on the part of the learners. In effect, an instructional style that grants more freedom and opportunity to the learner might induce better learning and a more positive attitude than a more controlled style of instruction. In these two learner control groups, students could not only control the content of learning, that is, the sequences of learning materials and examples, but also control the learning strategies such as the way of presentation and the quantity of practice. They could use the freedom which was provided through learner control and had to make their own learning decisions, therefore, spending more time learning than the program adaptive control group. In addition, while these students were progressing at their own pace, there was no doubt that these students could master all the content they had to learn. The results also showed that the variable of the extent of learner control has significantly interacted with learner's learning style. Field independent students performed better in the way of learner control, while field dependent students were better with program adaptive control and the middle students performed the best in the role of learner with advice control.

This study has explored CAI with a content of a moderate level of difficulty, mainly including conceptual learning. More research is needed to explore which extent of learner control under different levels of difficulties would be more effective.

Learning skills training and cognitive styles
Further exploration about the effects of different types of guidance on learning skills, cognitive styles, and the model of learner control have been conducted [7]. The task of this study was for students to learn how to use Chinese punctuation through CAI. Three groups of 143 students in total had been randomly selected from first year of junior high school. Each group was treated differently. Group A was learning Chinese punctuation following the procedure which was prescribed by the courseware itself. No tutorial of learning skills was involved in. Group B was learning through the courseware which had some tutorials built-in, that is, hints or cues for key words and learning skills. Group C, however, had teacher's guidance in addition to the courseware with

tutorial. This meant that the teacher taught one or two examples in the beginning and pointed out the focus, explained the difficult points and reminded students of the important feedback from the computer. Also, during the learning period, the teacher walked around the classroom, answered questions raised by the students, and provided some advice of learning methods.

Results showed that Group C, using courseware with tutorials combined with teacher's guidance, gained the best achievement, obtained the best effect on transfer, and used less time than others. Considerable interaction between the cognitive style of students and the transfer effect on CAI of the Chinese punctuation was indicated. The transfer effect of field independent students proved better than that of field dependent students, especially in their achievement of the post-test. The results were quite clear. After they had really grasped the propositions and rules about the Chinese punctuation, they were able to use them better; even though they were given a substantial problem in their learning situation, their success was quite consistent as well as the transfer effect. This demonstrated that they could take advantage of their inner reference to make their judgment independently. Group C and Group B were both involved with guidance on learning skills, with the implied elements of meta-cognition; hence their effect of transfer should be more conspicuous.

CONCLUSIONS

From these experimental research studies, we have not yet made an all-round and detailed analysis on the variables and parameters in the experimental program. We have so far only selected the findings on learning characteristics of students themselves and the teachers' guidance, in order to offer some facts. We want to show that these two variables are of great importance in their influences on the effectiveness of CAI. The special features of students include their cultural background, learning basis, learning style, motivation, and their expectation in the learning environment offered by CAI. All of these can have an influence on the learning effect. Students themselves are the *masters* in the learning process, also the *inner factor* of their development. No matter how advanced a technique, a method, or a measure is, only through the student who is the vital factor as the mediator in the learning environment, can such a resource realize its potential function. In this process, there's no exception, even with information technology and the computer as its core.

This point is what our experimental researches have indicated time and again.

Starting from John Dewey, some western educational theorists have emphasized the child-centred education in regard to the statement above. I agree with this stand-point. But I do not entirely agree with the child-centrists because the primary and secondary students are growing individuals. Whether in terms of their socialization or of their development in knowledge, intelligence, and technical ability, they have to experience the process from immaturity to maturity, from unknowing to knowing. "Learning by doing" is a quite good method as they learn from their direct experience, for it can meet the individual needs to learn at his/her own pace. However, it is neither possible nor necessary for them to learn by experiencing everything. For one of the special features of student learning is to learn mainly from indirect experiences, and here, *the teacher's guidance still plays a very important role*. The teacher's guidance can protect students from wasting time on "trial and error" in their learning process, help them to speed up their process of information processing and thus gradually raise the quality of this process. With all the above, our three studies have proved this point. Though it is not immediately clear in the first experiment that the teacher's guidance has a positive effect on the students' achievement, it can be seen definitely from the inquiry on attitude. In the other two studies, learner control with guidance and courseware guidance combined with teacher's guidance both had a positive effect.

Thus the "outer condition", especially the teacher, cannot be neglected. The inner factor is the base of learning, the outer factor the condition of learning, the outer factor functions through the inner. This is a famous saying by Mao Ze-dong [8]. It is ultimately appropriate to cite here to explain the relationship between child-development and education, the relationship between teaching and learning. It is natural that with the growing and maturing of psychological development, the teacher's guidance can be gradually simplified and reduced. But the appropriate amount and the extent of this guidance still needs to be explored. Convincingly, even adults need necessary guidance in the process of learning new knowledge and thus complete the leap from unknowing to knowing. According to Piaget's theory of cognitive development, some adults might not reach the formal operation stage even in the life long time. Therefore, the training on learning skill is not only beneficial to the learning process of students, but also advantageous to the adults' learning.

I believe that the integration of information technology into the instructional process develops an immense potential for the student's learning environment. However, the technology itself cannot automatically release its potential; only teachers who have mastered the student's learning patterns are aware of the conditions under which this new technology can play its role efficiently. Then a student's learning and development can be promoted consciously.

REFERENCES

1. (1990) *Encyclopedia of Chinese Education*, Chinese Encyclopedia Press.

2. Chen, Q. (1990) *Seven Years' Retrospect and Prospect: Computers in Chinese Secondary Education*. McDougall, A. and Dowling, C. (eds) Computers in Education: Proceedings of the 5th World Conference on Computers in Education. IFIP, Elsevier Science Pub. Co., North-Holland.

3. Chen, Q. editor (1987) *Readings On Computer Education in Secondary Schools*. GuangMing Daily Press.

4. Chen, Q. (1991) *Variables Moderating the Effect of Computer in Education*. Proceedings of IFIP WG3.1 Working Conference, UCSB. Elsevier Science Pub. Co., North-Holland

5. Chen, Q. (1992) *Effects of Computer Assisted English Learning on Students' Achievement and Attitude*. Proceedings of The Second Afro-Asia Psychological Congress, Beijing University Press.

6. Xia, W.F. & Chen, Q. (1991) *The Effects of Types of Instructional Control and Cognitive Styles on Tutorial CAI*. (in press)

7. Qing Z. J. & Chen, Q.(1992) *An Experimental Study on Learning Chinese punctuation through Computer Assisted Instruction*. (in press)

8. Mao, Z.D. (1943) *On Practice*. Selected Works of Mao ZeDong. People's Publishing House.

Qi Chen,   formerly Dean of Faculty of Education at the University of Macau, is now Professor of Educational Psychology in the Department of Psychology, Beijing Normal University and Deputy Director of the National Research Centre on Computer Education in Schools. She is also a member of the Board of Directors of the Chinese Psychological Society and vice president of the Educational Psychology Committee.   Qi Chen was a visiting scholar in Psychology and Education at UCLA during 1980-83 and has been doing research on educational computing for 10 years. She believes that with more understanding about student learning, information technology could be more effectively integrated into education.

19

Post-experimental phases of IT across the curriculum projects: the Spanish view

Carlos San José

Programa de NTIC, Ministerio de Educación y Ciencia
Madrid, Spain

ABSTRACT

This paper describes how the new information and communication technologies - mainly computers - have been introduced in secondary education in Spain. Eight different plans for the introduction of IT in education are being carried out at the moment in Spain, but this paper deals only with the ATENEA Project, managed by the Central Government's Ministry of Education. The main topics covered here include policy-making concerning software and hardware, teacher training and Telematics. Some remarks, relevant from the point of view of a well implemented, post experimental Plan, have also been included regarding the main issues involved in the educational use of IT and its integration in the curriculum.

Keywords: management, national policies, teacher training, communication

INTRODUCTION

This paper describes how the new information and communication technologies - mainly computers - have been introduced in the structure of Spanish secondary education.

The structure of Government and Administration in Spain is becoming territorially decentralized through a process that affects every sphere of social life, including education. Jurisdiction on educational matters has already been transferred to the local authorities of seven of the seventeen Autonomous Regions (Comunidades Autsnomas), but in the other ten - roughly speaking, half the territory of Spain - education is still managed but the Central Government's Ministry of Education. For that area, the Ministry created a department called Program for New Information and Communication Technologies (PNTIC), which is in charge of the introduction of IT in the schools (i.e. education below university level), and has developed a plan called ATENEA Project.

Similar projects have been implemented in the seven Regions with full jurisdiction over education, but the ATENEA Project is the most important of all the national plans in terms of size. In spite of their differences, the eight plans share some common features. In order to achieve cooperation and convergence, the heads of the projects hold frequent meetings, chaired by the Director of the PNTIC and the ATENEA Project.

THE ATENEA PROJECT

Educational projects which are innovative both in their contents and their context cannot be implemented by national ministries or departments of education without a prior assessment of their global impact on the whole educational system. It is essential not to neglect the impact of IT projects on regulated activities (curriculum) and school management and organization (classroom space, timetables), or the various factors involved in teacher training, to mention but a few of the main key issues.

In operational terms, the feasibility of such projects depends on policy-making at three levels: hardware, software and teacher training. Projects can be judged according to way these three critical factors are dealt with.

From this point of view, comparisons between the different member states of the European Union [1], and even between different nations in other areas of the world [2], show that approaches are very similar in most countries, the only important differences arising from issues concerning centralization or decentralization.

Hardware policy
The first of the above mentioned critical levels of policy-making comprises the most conspicuous aspect of any Project: choice of standards, big investment planning, site enhancement, contracts on hardware maintenance, design of the distribution network, etc. In the Atenea Project, decisions on the provision of equipment are centralized due to purchasing costs, although many public schools may have alternative means to provide themselves with equipment.

In this respect, centralization has many advantages over decentralization, which implies in practical terms that schools or small groups of schools have to buy their own hardware. Strength in big numbers, both at the initial purchase and for the subsequent maintenance scheme, is a key factor.

Software policy
It is important to have a clear policy in this field, due both to linguistic-cultural implications and issues concerning national industries. When the first projects were launched in Spain, around 1985, there was hardly any software available in the Spanish market which could qualify as "educational". The solution that was then adopted was to resort to general purpose software and explore its educational potential. Some integrated packages were (and still are) used to develop the pedagogical use of word processors and data bases. Spreadsheets were less frequently used, due to their complexity and difficulty of use.

At present a remarkable amount of educational software is available from different sources, always in national versions.

Regarding software, language is a delicate issue in any country where English is not the native tongue. Taking into account the age of potential users, and their non-professional (or pre-professional) condition, the Spanish Ministry of Education decided from the start that all the software used in schools had to comply with the sine qua non requisite of being in Spanish. The Spanish-speaking market is big enough all over the world (300 millions of native speakers) so as to lure both big and small firms into producing Spanish versions of their programs. This requisite has been seriously fulfilled.

Two other initiatives of even greater importance have also been implemented in this area, in order to encourage the creation of a national educational software market, with Spanish characteristics. The Ministries of education and Industry signed an agreement to jointly provide funds for R+D projects, an initiative which has helped to fix standards (concerning interfaces, prices, distribution networks, etc.) in the Spanish market. On the other hand, public software contests have been organized periodically, offering substantial awards. About one hundred good quality programs have been obtained from this source, their copyrights now belonging to the Ministry of Education.

Teacher Training

This is probably the most important issue. Hardware affairs are largely a money matter, and software issues, being more complex, can be solved with the effort of a relatively small team. But the decisive factor on which the success of the whole plan depends is teacher training. This fact is widely recognised in most of the international literature [1],[2] and [4].

In Spain, a pre-existing network of Teachers Centres was enhanced and used to provide training in the educational use of IT. Plans for teacher training are developed at two levels. Some teachers are temporarily relieved of their normal duties in order to receive intensive training and a specialized instruction, which enables them to impart seminars and provide advice to other teachers. There is a vast network (113) of Teachers Centres at the moment, there being one of these IT specialists in each and every one of them. The rest of the teachers involved in IT activities at their schools may receive general training outside their normal working hours, as they are not relieved of their duties.

In every school participating in the Atenea Project there is a Supervisor and a teacher in charge of co-ordinating IT activities and providing a link with the Teachers Centre. The Headquarters of the Project provides the training of the supervisors (150 hours) on IT and methodology. The rest of the teachers in the school team receive their basic training at the local Teacher Centres.

Finally, it is to be mentioned that the network of Teachers' Centres seems to be quite capable of catering for the training needs of private or public sector teachers interested in IT. Nevertheless, the negative fact that training cannot be imparted within normal teaching hours, which seriously hinders its expansion to a massive scale, will have to be confronted in the future. The most fruitful policy in this respect would be to include IT in pre-service training. In the meantime, Telematics and distance training will have to play a more decisive role in teacher training, as is showed in the experiments already done.

Objectives
The objectives of the ATENEA Project may be summarized as follows.

Focussing on the students
- To favour cognitive development and innovative learning by means of new environments.
- To stimulate critical understanding and rational uses of new technologies as means of expression.
- To enable students to access, organize and process information by means of new technologies.

Focussing on the teachers
- To provide the technical support and the proper training which will enable them to use computers as pedagogical tools and instruments for innovation and improvement.
- To enable them to select and analyze the resources best suited to their environment and their specific tasks.
- To improve the management and organization of schools.

Focussing on the Curriculum
- To establish patterns for the integration of NIT in the different curricular areas.
- To enhance the influence of computer science and information technology in the curriculums of all types of general and specialized instruction.

Equal Opportunities
Equality of opportunities for women has been considered of paramount importance. Special courses for female teachers have been organized with the aim of encouraging projects launched by women, and counterbalance the male monopoly of IT, both as regards teachers and students. However, as the problem has been detected mainly in the teaching profession, measures have been taken mainly to provide female teachers with equal opportunities. On the other hand, although there does not seem to be a noticeable male predominance in the use of IT by students, more careful studies are being carried out in order to establish whether the general technological orientation of males may be mirrored in the field of IT in the future.

Other actions are being carried out for students with special needs and underprivileged social groups. New technologies have proved to be useful tools for the integration of these groups, and the future trend is toward increasing actions in this field (specific training for teachers, production of materials, provision of equipment to schools).

The evaluation of the ATENEA Project

The ATENEA Project was evaluated by the OECD [3] in 1990. This process was the final stage of the so-called pilot phase of the Project (1985-1990). The evaluation team analyzed every dimension of the Project and elaborated a set of considerations and warnings that have been the basis for the Project's continuity into the extension phase. The Project is reaching at the moment its last stage - the generalization phase. This is the natural next step and will provide every Spanish secondary and primary public school with access to the facilities, within the framework of the educational system defined by the new legislation recently introduced (1992).

Some of the problems detected by the evaluators were: the lack of pre-existing materials developed for an innovative project, and the slow pace of implementation of the whole Project, mainly due to the amount of time necessary for software development and the slowness and complexity of the purchasing mechanisms after development. The resulting lack of materials has recently become a serious problem.

Problems were also detected in the teacher training process, namely the time span (a whole school term) required for training a specialist, the fact that school teachers received training only on a voluntary basis (outside working hours, and with no supplement to their salaries); and the de-motivating extra work involved in changing habits when using new tools. Problems of school management (organization of space, computer classrooms, security measures, maintenance costs) were also found to be of some importance. Many of these problems are being addressed by the new Education Act, which provides the schools with some independent jurisdiction over curricular design and adaptation as well as timetables and organization schemes.

The OECD Report also touched upon the schemes for the provision of equipment and the horizontal communication between teachers. In the evaluator's opinion, the schools should play a more active role in the provision of their own equipment, which would enable them to become more independent from the centralized policies of the Ministry. This would help to diversify the equipment and would contribute to solve the problems arising from the obsolescence of materials.

TELEMATICS IN EDUCATION: THE PLATEA PLAN

Following the international trends in the field of educational Telematics [5], and the advice of the OECD Evaluation Report [3], the PNTIC has organised different telematic infrastructures and activities to consolidate the links between members of the educational community. The success in this field is one of the indicators of the Project being in a full blossoming phase.

The school Telematics plan is called Platea. The objective of this project is to provide teachers with facilities for communicating, exchanging and disseminating their experiences easily. As such, Platea is also a medium for teacher training. It encourages cooperative work between teachers and pupils in different Spanish or foreign schools, by means of the sharing of experiences and the preparation of joint work. Finally, the Platea plan is trying to help students incorporate the telematic tools in their work.

Two types of networks have been initially created: a central network, and several local networks. The central network, whose node is located in the PNTIC headquarters, is based on the Spanish videotex network, Ibertex. It consists of a Service Centre specifically created for schools, teacher training centres, and academic authorities, although most of its services can be accessed by whoever uses Ibertex. The main advantage of this solution is the low fares scheme (about 3,5 ECU per hour, regardless of distance). The main disadvantage lies in the limitations of the resources (speed, facilities) provided by the network.

The second kind of network is a net of local networks based on the Telephone Switched Network (Red Telefsnica Conmutada, RTC). Each of the 113 Teacher Centres becomes a regional node which can be accessed by means of a local telephone call. These networks can adopt the configuration of a central network, since all users can connect themselves to the PNTIC's node, but the cost of a phone call from outside the local area is too high for a normal school's budget. In the future, depending on negotiations with the Spanish PTT, it could be possible to connect the schools through the RTC network. The next step will be to provide an INTERNET access for every school thus opening them to the broad space of world communications.

THE MENTOR PROJECT: TELEMATICS FOR DISTANCE LEARNING

The PNTIC is also in charge of the Mentor Project, an experimental plan for open education and distance training for adults by means of telecommunications technology. One of its objectives is to test out these tools and the methodology for their use in education. Training and local socio-economic development are its basic aims, with a   perspective entirely different from that of other initiatives developed by the PNTIC.

A description of the Mentor Project would be beyond the scope of this document; suffice it to say, however, that it aims to develop open education and training in the less economically successful regions of Spain, and especially in rural areas where access to information and training resources is still a serious problem, in order to improve general culture among young and adult inhabitants of these areas, and to favour communication and access to information.

SOME REMARKS ON IT IN EDUCATION

Today's citizens must incorporate IT skills to their basic training. The use of certain tools, such as word processors or data bases, has become so unavoidable in intellectual activities of all types, that not using these resources in education would result in a waste of opportunities for enhancing human intellectual potential.

But the situation is far from being static or closed: new software and hardware is continuously being added to the list of indispensable (or at least desirable) educational materials. These tools allow us to perform otherwise impossible tasks, or simplify those which can be carried out by other means, increasing efficiency and reliability.

IT, in contrast with most sectors of trade and industry, shows a steady trend toward decreasing costs, in hardware as much as in software. This factor may allow schools, usually operating on very tight budgets, to access the latest available technologies that may prove to have some educational potential. Continuous investment is therefore required in both hardware and software, as well as in teacher training, which is the most important.

The concept of permanent training is of special importance, as teachers must remain both competent users of IT and competent instructors of the subject matter in which they are specialized. Methodologies, curriculums and school organization change over time, and teachers must keep up to date with these changes, some of which are themselves brought about by IT.

The integration of information technologies in the curriculum

Existing studies [4] have shown a strong trend towards integration of basic tools in the curriculum on a cross-curricular basis. This can be seen as the "normal" penetration of technology in the specific field of education.

But IT has also proved its efficiency in certain fields of activity which are now hardly conceivable without them. Such is the case of Special Education, where IT operates as an empowering prostheses, facilitating action which in many cases would otherwise be impossible. In the field of drawing and graphic design, professional activities have undergone such sweeping changes that they are almost exclusively based on the use of IT at the moment - perhaps with the partial exception of the area of artistic creation. The same holds true for education: any curriculum attempting to be complete must include the use of computer aided design and drawing, not because the basic skills required for handling paper, pencil and ink should be discarded, but because there is a whole new horizon of possibilities for experimenting and learning now open even to the non-gifted.

A similar case can be found in the field of the automatisation of technology. There is a kind of technology based on handicraft techniques, and another more advanced type which resorts to IT.

The school newspaper, perhaps the most popular of cross-disciplinary school activities, can also be transformed by the use of IT. This is a very enriching activity which allows for topics of social, literary, or artistic interest to be dealt with, providing at the same time opportunities for the practice of critical and writing skills, and for the development of cooperation between the members of the school community. ITs not only simplify the tasks involved, they also add new dimensions to the pedagogical experience, such us desktop publishing, word processing, Telematics. Other educational areas where the use of ITs is expanding are physics and chemistry laboratories and computer aided musical instruction.

CONCLUSION: TRENDS

One of the problems we face at the moment is the high costs of hardware in comparison with other types of school equipment, an obstacle which seriously hinders its incorporation to classroom practice. In the Spanish public sector, the only solutions considered for the time being are based on centralized purchase of equipment. In the future, as costs are reduced and computers and peripherals become widespread in different social areas, local institutions will play an important role as equipment providers, and normal school budgets will contemplate the allocation of funds for IT materials less exceptionally than they do now. In the private sector, funds will be obtained from the usual sources, and the pace of development will be that established by the law regulating the educational system, and by the example set by the public sector.

With regard to the use of technology in the classroom, three trends are foreseeable. Firstly there will be an increase in curriculum-integrated use, which may rely on higher quality, more specialized and powerful software such as multimedia and LANs. The aim will be to teach skills and concepts by means of new technologies wherever they prove more useful than traditional resources.

Secondly, IT will obviously be used to deal with certain IT-based areas of knowledge, such as CAD, Robotics, infographics, laboratories and microelectronics. The increasingly less important field of programming languages could be included here. Thirdly, computers will be used through word processing, data bases and CD ROM drives for production , access to information, and the management of the teaching and learning process.

REFERENCES

1. There are 12 Member States Reports on New Information Technologies in Education, elaborated for the Commission of the European Communities-Task Force of Human Resources and published by the Office for Official Publications of the European Communities, Luxembourg (1993): *New Information Technologies in Education in*: *Belgium, Denmark, France, Germany, Greece, Ireland, Italy, Luxembourg, The Netherlands, Portugal, Spain, The United Kingdom.* OECD, Luxembourg.

2. Pelgrum, W.J, Jansen, I.A.M. & Plomp, Tj. (1993) *Schools, Teachers, Students and Computers: a Cross-National Perspective.* IEA. Enschede.

3. Ministry of Education and Science (1991) *Proyecto ATENEA: OECD Evaluation Report.* Secretaría de Estado de Educación, Madrid.

4. Ministry of Education and Science (1991) *Information Technology in the curricula of the different E.C. Countries.* Secretaría de Estado de Educación. Madrid.

5. Veen, W and Vogelzang, F. (1992) *Telematics in Dutch Education.* Academisch Boeken Centrum, De Lier.

Carlos San José Villacorta is Technical Councillor of the Program for New Information and Communication Technologies (PNTIC), Ministry of Education and Science. After a period as a teacher of mathematics in secondary school, he is now a tenured civil servant in the Systems and Information Technology High Corps, in charge of national and international projects dealing with development, applications and training. Carlos is a member of several directive boards and working groups in European projects dealing with educational software, teacher training and ITC, and computers in education; he also serves on the expertsd panel of the Telematics for Education and Training project of the European Commission. He has written several books and essays on computers in education.

20

Social and political influences on the integration of informatics into Japanese education

Takashi Sakamoto

National Center for University Entrance Examination
Tokyo, Japan

Keiko Miyashita

Tokyo Institute of Technology
Yokohama, Japan

ABSTRACT

Government and other segments of society have been taking a leadership role in integrating informatics into Japanese education. Reports of educational councils, reports of task forces, consigned research studies, financial support, teacher training, catalogues for educational software, and other support provided by the government have influenced the integration of informatics into Japanese education in a positive manner. This paper address the brief history of Japanese computer education, how computers are integrated into the school curriculum and school subjects, and current governmental initiatives.

Keywords: government, national policies, curriculum development, information society

INTRODUCTION

Information technology has been rapidly developing. Especially noticeable is the fact that computer hardware has been improving every day. It might be said that today's technology is no longer useful for tomorrow. People have to educate themselves using appropriate information technology in order to actively cope with the dramatically changing modern society as well as to participate in changing society itself. In Japan, educators have been trying to foster students' self-teaching abilities and information literacy. These abilities will be needed by the students in the future when they become adult citizens.

HISTORY OF JAPANESE EDUCATIONAL COMPUTING

Philosophical background

The Ministry of Education, Science and Culture of Japan (Monbusho) has taken a leadership role in guiding the use of computers in Japanese schools. Several different divisions of Monbusho are now related to Japanese educational computing. In 1985, the subcommittee on Educational Media of the Social Education Council published "The Report on the Use of Microcomputers in Education," and this report became the general guidelines for the introduction of computers into schools [1].

The National Council on Educational Reform produced four reports (1985, 1986, 1987 and 1987) which laid the foundation for the rapid introduction of information technology into Japanese schools during the 1990s [2]. In the first report, "to cope with informationalization" was identified as one of eight important issues. In the second report, three principles for informationalization and the need for teaching information literacy were mentioned. In the third report, a plan for intelligent schools was discussed, and in the final report, internationalization and informationalization were mentioned as the most important issues of education. The three principles for informationalization noted in the second report [3] were:

- to promote eagerly the cultivation of information literacy;
- to utilize the potential of information technology in all educational institutions; and
- to reduce the negative effects and increase the benefits of information technology.

Corresponding with the work of these councils, the Task Force on Elementary and Secondary Education in the Information Society also played a major role in developing Japanese computer education. In the first report which was published in 1985 [4], the basic philosophy of how computers should be introduced into different levels of the schools was noted:

- *Primary school*: To familiarize pupils with computers thorough their use as teaching tools;
- *Lower secondary school:* To help students acquire computer awareness and literacy by making greater use of computer functions such as simulation and information retricval.
- *Upper Secondary School:* Special consideration to the progress of the information society and the effects of computers on individuals and society.

Thcse reports have been influencing the development of Japanese educational computing today.

EDUCATIONAL COMPUTING TODAY

Computers in the school curriculum
In 1987, the Curriculum Council published the final report on "The Improvement of Standards for the Curriculum for Kindergarten, Elementary, Lower Secondary and Upper Secondary Schools." [5] Two recommendations in the report were:

- to emphasize the development of self learning motivation and competence in active adaptation to society; and
- to consider the development of fundamental abilities necessary for adapting progress of science and technology to information.

The second recommendation was not included in the interim report which was published in 1986. However, since the importance of active adaptation to informationalization was discussed during 1986, the sentence, "to consider the developmental levels of students and appropriate guidance in dealing with computers" was added and became the foundation of the Standard Course of Study.

In 1989, the Standard Course of Study emphasized incorporating educational information. The Task Force on Elementary and Secondary Education in Information Society influenced the Standard Course of Study [6]. The Curriculum Council mentioned that "information literacy " were not only the abilities to use computers but also the abilities to judge, to select , and to process information. Moreover, the Council suggested that several important abilities such as creating new information, communicating using information, understanding the importance of information in society, understanding the importance of information property, and obtaining responsibility toward information, should be taught.

In 1993, "Fundamentals of Informatics" was added to the lower secondary school curriculum as an option in Technology and Home Economics, which is one of the compulsory subjects. "Developmental processes of computers" was added as an option area of Science and "Computers and Tiredness " was added to Health Study in the lower secondary school curriculum. At the upper secondary school level, the use of computers was encouraged in the curriculum of Science, Mathematics and Home Economics. Moreover, computers have been integrated into several school subjects, such as Social Studies, Geography, Political Science, Physics, Mathematics, and Music.

Fundamentals of informatics
As previously mentioned, "Fundamentals of Informatics" was added to the lower secondary school curriculum. Students learn computer systems, computer programming, utilization of hardware and software, social influences of computers, and the importance of computers.

First, students learn how computers are related to their daily lives, and are motivated to learn about computers. Regarding understanding computer systems, students learn the basic structure of computer systems and the function of software. In the area of computer programming, first, the students learn basic operations of computers. They learn how they can turn on the switches of each device, start and reset the computers, and use floppy discs. Then they learn the use of a keyboard. Since Japanese use several different alphabets, the students can input words using either the English alphabet or Japanese alphabets. Then students try to use several basic applications such as Japanese word processing software, database software, spreadsheet software, and graphics software. The objective of using these applications is that the students should be able to select information, arrange information, manage information and express their thoughts. The students also learn introductory programming, such as a programming language, BASIC.

Finally, the students learn the role of computers in industry and in their daily lives, and social influences of computers. This part is strongly related to the third principle for informationlization, "to reduce negative effects and increase benefits of information technology," which is in the second report of the National Council on Educational Reform. The students learn a brief history of computer developments. Then they learn several examples of the use of computers in their daily life. For example, the use of computers in banking systems and reservation systems for transportation are discussed. The students learn the importance of computer networks. They also learn that many industrial designs such as automobile parts and bodies are developed with computers, and that the distribution system of our daily goods has been improved with the use of computers. Re-grading negative effects of information technology, the students learn some health problems and some crimes which are related to the use of computers. They learn the value of information and the importance of managing information. The copyright of software, illegal coping of software, and computer viruses are discussed. Students are encouraged to use information appropriately in order to develop a future information society.

A new concept for achievement
According to the new Standard Course of Study," in order to improve educational activities, educators need to help students develop their motivation for self learning and their abilities to actively cope with s changing society, to teach basic fundamental content, and to make an effort to deal with the students' individual characteristics." This sentence is related to objectives in each subject area of the Standard Course of Study. These objectives are to develop interest, motivation, and positive attitudes toward the subject areas, and to foster thinking skills, decision making skills, and skills of expression.

Japanese education has been moving away from behaviourism toward constructivism. Computers in education had also mainly been used with behavioristic theory. Therefore, most research studies were conducted in order to develop drill and practice, and tutorial types of CAI, and they contributed to the learner's acquisition of basic knowledge and skills using computers. The results of these research studies were important and influenced computer education in Japan. However, according to the constructivistic point of view, it is desirable for the learner to think, decide, and express by themselves. Therefore, today computers are being used as tools for student expression, for problem solving, for discovery learning, and for investigating.

Computers as educational tools

Formerly, students developed their knowledge acquisition abilities through computer experiences. In addition, today's students also develop their creative expression abilities through computer experiences. Computers are now being used to help students develop self learning abilities, self expression abilities, and problem solving abilities.

With respect to the Standard Course of Study, computers have been integrated into many school subjects. The Japan Association for Promotion of Education Technology published "Ideas for Using Computers in the Classroom." [7, 8] Several examples from different schools were gathered for these books. Some examples are discussed in the following section.

Science

Computers are widely used for science experiments. Students measure the conditions of nature using sensors, investigate a principle of nature using graphics and charts of these data, investigate rules of nature using simulations, and solve problems collaboratively using information in a data base. For example, in one lower secondary school, the students experimented with the boiling point and the melting point of several materials using a thermal sensor which is connected to a computer. Since the data for the experiments was presented in graphs by the computer, the students could concentrate on the change in the data. Then they could discuss their thoughts about the data. Other experiments, which are conducted for investigating the changes in temperature due to expansion and compression, and for investigating the body temperature of different types of animals, are also carried out with computers and sensors. Data base software is used for studying about plants, animals, rocks and fossils. In one lower secondary school, the students made a field trip to research rocks and fossils. They used a data base in order to classify the kind of rocks and fossils which were found, and made reports using computers. Computer simulations are widely used for the study of planets and stars.

Social studies

Students learn historical issues using a database, social systems using game type software, and discuss some social problems using multimedia. For example, the students in a lower secondary school worked on a group project about the history of the Meiji Era in Japan using a data base. They could search the necessary items and data in a short time. The students in another school used a computer simulation game in order to understand the role of the mayor and members of a city council, the administration of the city, and elections. The students became a mayor during the simulation game called TOKIO. As a mayor, the students considered many issues and tried to be re-elected at the end of the game. They developed a high level of problem solving skill during the activity. Multimedia was also effectively used in one school. The topic for the activity was Japanese car export problems with the United States. The students developed multimedia using several different types of sources, such as video tapes of TV news, graphics, newspaper articles, and so on. After making the multimedia screens, the students used the material for developing their opinions about the topic, and discussed these in the classroom.

Other subjects
Computers have been integrated into several other subject areas such as Mathematics, Music, Art, English, and Technology and Home Economics. The students in one school studied geometry and functions using simulation and graphics. At one school, the students composed music using computers and played some instruments with computers. At several schools, computers were used as a presentation tool. The students typed using word processing software and made spread sheets and graphics. Some schools encouraged the students to communicate with the students in other countries using computers. Students in these activities were motivated to study English and learned many things from friends in different countries. The students in one upper secondary school made a questionnaire and sent it to schools in foreign countries using the computer. The information from different countries helped them to discuss the similarities and the differences in each culture. For Ecological issues, computers were used in several schools. At one school, the students researched separate topics and input their data into a computer. Then they used other students' data in a data base to understand some other ecological issues. In another school, the computer simulation game BALANCE OF THE PLANET was also used for the study about ecological issues. After playing the game, the students in the class discussed the issues.

Computers became powerful educational tools in many schools. In summary, computers have been used to help students develop self learning abilities, self expression abilities, and problem solving abilities.

TODAY'S ISSUES

Some problems in schools

In the earlier period of introducing computers into schools, some of the experienced teachers were unwilling to use them for class teaching. But now where there are many computers, these teachers now make use of them.

Concerning hardware diffusion, for the last five years, the governmental budget had supported 22 units for all primary schools and 42 units for all secondary schools. One third of the budget for matching funds came from the ministry of education and another two thirds should have been paid by block grants from the ministry of local affairs to local authorities. One of the big problems has been a failure to understand about the availability of this budget from block funds. Many educators in the local districts usually do not know about this system for getting computers. As they do not make demands from this budget, in many districts much of the money for computers has been directed instead to the construction of bridges, roads and buildings. In order to improve this situation JSPET has published and distributed a booklet describing this budget system to local educators, encouraging them to use it. But from this year, all expenses will be provided by block funds for a lease/rental system.

Another problem is the lack of an infrastructure for supporting initiatives in learning. Ordinary schools have only two or three-telephone lines and one computer room. There are few computers in science laboratories, music classrooms, fine art rooms or other special areas. So teachers and students experience difficulties using computers as tools for creative study and communication.

Concerning software, the amount of good commercial software has increased over these ten years. In the earlier stages, many teachers developed and used their own software. The ratio of teacher-made software to commercial software in lower secondary schools was 37% to 23% in 1985, but it became 6% to 86% in 1992. At the moment, the average lower secondary school has 33 kinds of software and 200 pieces of software. This figure shows the number is still small and teachers cannot always use suitable software.

Concerning teachers, one third can use computers, but only 10% can teach students to use computers in their schools. In the pre-service teacher training, there is not yet enough computer education. Many students in the teacher training courses in ordinary universities study only a few hours on educational methodology including computer use, as a compulsory subject. Some of them just watch video programmes of school children using computers in their classrooms.

Among enthusiastic teachers some only use computers for all kinds of presentation which could be done more efficiently by a different media. Finally, in most schools, teachers usually do not allocate a budget for upgrading software versions, for buying new software or even for maintenance. As a result they often still have the same configuration of hardware and software as they did in the first introductory phase.

CURRENT GOVERNMENT INITIATIVES

The age of "ubiquitous computing " is just beginning. This concept comes from the fact that everybody will be able to use computers everywhere. The technology which leads the age of "ubiquitous computing" is the work station, down-sized personal computers, networking by telecommunication, and multimedia.

Today, computers are being widely used in business fields in Japan; however, the use of the computer for planning, management, and administration is less widespread than in the other countries. In 1993, the Information Industry Committee of the Industrial Structure Council which is in the Ministry of International Trade and Industry, (MITI or Tsusansho) reported on the analysis of the present situation and suggested ways to improve the current situation. At the same time, the sub-committee for Development of Human Resources in an Information-Oriented Society, part of the Information Industry Committee, suggested the necessity of developing several types of information engineers and the curricula to educate them in order to improve the quality of system engineers [9]. According to this report, there should be 13 different curricula for educating information engineers, 2 common curricula, and 2 curricula for the users. These different types of information engineers are:

- system analyst,
- project manager,
- application engineer,
- production engineer,
- technical specialist

- system use and management engineer,
- development engineer,
- system administrator, and
- educational engineer.

Educational engineering was pointed out as a profession, and the curriculum and the qualifying examination are being developed now. The skills which are needed to become an education engineer are:

- skill for making another person's teaching plan,
- skill for educating knowledge and techniques to others,
- skill for making presentations with technical writing,
- skill for developing high quality educational materials such as multi-media materials using information technology.

It is a remarkable development that educational engineers will be included with information engineers in those certificated by the Ministry of International Trade and Industry.

The other major governmental movement today is toward multimedia. Many research studies have been conducted about multimedia. The Japan Audio-Visual Education Association has been developing "Science Hyper Media" as a teaching material. Moreover, the Japanese Educational Materials Research Institute has been exploring appropriate uses for multimedia type data-bases. The Research Institute for Software Engineering also made a report on a new informational education system. The new system is game type English education material. The Center for Educational Computing has been developing several educational software products, such as software which uses multimedia, expert systems, and simulations, intelligent adventure software for the music world, sound adventure software, software which deals with energy and ecology, and many other types of software. As this research which is related to multimedia is becoming further developed, it is necessary to have specialists who produce the multimedia software. The International Multimedia Association conducted research on the usage of hardware, diffusion rates for software, consumption and development of software, and the present status of multimedia development specialists. It suggested the importance of promoting and educating many specialists.

Research studies related to The Ministry of Post and Telecommunication (Yuseisho) have been conducted recently. The Telecommunication Advancement Cooperation published a report [10] suggesting that educators needed to teach information technology, including telecommunication and broadcasting, appropriately. It also mentioned the importance of developing a policy which dealt with networks for telecommunication and broadcasting, since informationalization of education has been moving toward the use of networks. Moreover, the International Communication Fund has been conducting research on international distance education and conducting pilot research studies.

CONCLUSION

As shown in the Figure below, government and other segments of society have been taking a leadership role in integrating informatics into Japanese education. Reports of educational councils, reports of task forces, consigned research studies, financial support, teacher training, catalogues for educational software, and other support provided by the government have influenced the integration of informatics into Japanese education in a positive manner.

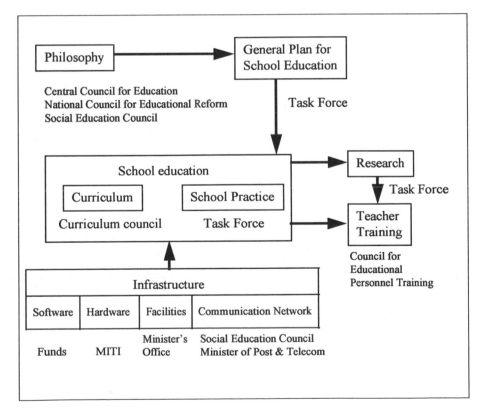

Figure 1: Leadership roles in integrating informatics in Japan

REFERENCES

1. Monbusho Social Education Council (1985) *The report on the use of microcomputers in education* (Japanese).

2. Knezek, G. & Sakamoto, T. (1991) *Teacher training for information technology in Japan and the united states.* Proceedings of ICOMMET '91 Tokyo, Japan. 297-300.

3. National Council on Educational Reform (1986) *The second report on educational reform.* Ministry of Education, Science and Culture of Japan.

4. Task Force on Elementary and Secondary Education in the Information Society (1985) *The first report on the task force on elementary and secondary education in the information society.* Elementary and Secondary Education Bureau. Ministry of Education, Science and Culture of Japan (Japanese).

5. Monbusho Curriculum Council (1987) *The improvement of standards for curriculum for kindergarten, elementary, lower secondary and upper secondary schools* (Japanese).

6. Monbusho (1989) *Standard course of study.* Ministry of Education, Science and Culture of Japan (Japanese).

7. Japan Association for Promotion of Education Technology (1992) *Ideas for using computers in the classroom. Vol 1* (Japanese).

8. Japan Association for Promotion of Education Technology (1993) *Ideas for using computers in the classroom. Vol 2* (Japanese).

9. Subcommittee for Development of Human Resources in an Information-Oriented Society (1993) *The final report of the subcommittee for development of human resources in an information-oriented society.* Ministry of International Trade and Industry (Japanese).

10. Telecommunication Advancement Cooperation (1993) *The plan for information environment and information literacy in the information society* (Japanese).

Takashi Sakamoto is Vice President of the National Centre for University Entrance Examinations. He is also Emeritus Professor at the Tokyo Institue of Technology, Japan. His background is Psychology of Thinking and Learning and Educational Technology.

Keiko Miyashita is a JSPS Post Doctoral Fellow in the Department of Systems Science at the Tokyo Institute of Technology, Japan.

21

The failure of the education system to attract girls to formal informatics studies

Joy Teague
Valerie Clarke

Deakin University
Geelong, Victoria, Australia

ABSTRACT

Although computers are widely available in the workplace, the school room and the home, statistics indicate that informatics still appeals more to males than to females. Research has shown consistently that boys utilise computers far more than girls in the home, at school and in the community, and receive greater encouragement for doing so. In this paper some historical reasons for this are given, the current situation in Australia is defined, attitudes of Australian schoolgirls to informatics are addressed, some recent educational initiatives described, and suggestions for increasing schoolgirl interest and participation in informatics are made.

Keywords: social issues, equity issues, attitudes

INTRODUCTION

Observations or surveys of enrolments in secondary and tertiary level computing classes, participants in computing games arcades, entrants and winners of computer programming contestants, participants in computer holiday programs, unscheduled extra-curricula users of school computing facilities, and owners and users of home computers consistently indicate that males have a greater interest in, and greater access to, computers than do females [1]. However, studies of abilities show that there are minimal gender differences in general ability, computing ability or computing achievement, and that, when girls are required to enrol in compulsory courses in informatics, they perform as well as their male colleagues despite their more limited experience on entry to these courses [2]. Thus gender differences in interests and participation in informatics seem to arise from the way in which informatics is introduced into schools [3], a process which facilitates the development of gender differences in interest in informatics and beliefs about the gender appropriateness of informatics studies and careers.

An historical perspective
It is interesting to note that in the early days of computers these gender differences did not exist. Two women who were well known for their very early involvement with computers were Lady Ada Lovelace and Admiral Grace Hopper. In the very early days of computers women were employed as programmers as most young men were away at war. It has been suggested that women were preferred for programming work because they had the necessary "patience" and "attention to detail" for what was regarded as an essentially low-level clerical job [4].

In the 1960s entry into informatics occupations was by way of "aptitude tests", generally a type of intelligence test. As women and men, from arts or science backgrounds, are equally likely to score well on such tests, women were as likely as men to be employed. Women working in informatics in the early 1960's recollect that the field was not viewed as being male dominated at that time. It was the developments which occurred over the next twenty years which have established computers as being primarily the province of men and boys. These developments occurred in two periods: the early 1970's and the mid-1980's.

Around 1970 computer science was established as a university discipline located in science and engineering faculties. As the normal prerequisites for science and engineering courses, science and mathematics, have traditionally low female enrolments in most Western countries, relatively few women enrolled in computer science courses. These courses became the standard entry requirement for informatics occupations, and the profession became stereotyped as a male occupation. Consistent with social role theory, girls perceived the gender imbalance in informatics courses and occupations and made the incorrect assumptions that existing jobs reflected existing abilities and attitudes [5], or that men had the interests, attitudes and abilities required for courses in maths, science and computing and that women lacked these abilities, interests and attitudes. Such perceptions influenced their beliefs of the suitability of informatics, not only for men and women in general, but as subjects and careers they should consider for themselves.

Around 1980 mini- and micro-computers became affordable and widely available. These computers were introduced initially into schools by mathematics and science teachers, with computing classes reflecting teachers' interests in mathematics and science. Software was limited. Most computers were purchased with 'on-board BASIC', and users were expected to write their own programs. Computer use required skills with complicated operating systems and editors, which limited computer usage to those who had the time and the interest to spend learning these systems. For societal and cultural reasons, these users were more likely to be male than female [3, 6]. At about this time, tertiary level female computer science enrolments in Britain and the United States began to drop, and declined steadily throughout the 1980s [7, 8].

Hardware and software developments
The problems arising from a shortage of appropriate software have been resolved. Today there is an extensive and diverse range of software available, enabling computers to be used with little or no formal training. The majority of programs are menu driven and contain explicit directions enabling them to be used with only an hour or two of training in computer use. Costs of computers have declined dramatically, placing them within the reach of a greater section of the population. Most schools have at least one well equipped computing laboratory. Many homes have one or more computers. Schoolchildren know that most members of the workforce use computers in their jobs.

What has been the effect of this hardware and software revolution on schools? There has been a realisation that computers can be used across the curriculum. Word processing, spreadsheets, and graphics improve the quality of most assignments. The teaching of humanities subjects, such as history and economics, can be enhanced with database packages. Specialised teaching packages are available in most disciplines. The world of multimedia is developing rapidly.

At informatics education conferences teachers and developers discuss innovative and exciting applications of computers across the curriculum. Attending conferences, such as the annual United States National Educational Computing Conference or the Australian Educational Computing Conference, makes one aware of the diverse range of available software, the broad penetration of informatics into the school curriculum, and the remarkable achievements of dedicated and enthusiastic teachers. With the hardware and software available today, informatics should be available to all students in all disciplines in all schools. However, this does not seem to be the case.

The current situation in most Australian schools
The reality of computer use in Australian schools was clearly explained by a regional school computing co-ordinator, who described the situation in his own area, but indicated that this was relatively typical of most regions. This co-ordinator is responsible for providing advice, information and short courses to teachers in more than one hundred primary and secondary schools spread over some twenty thousand square kilometres. Very few schools (none in his region) have sufficient resources to allocate one computer to each student or pair of students, so computers are placed in laboratories with selected classes being scheduled in these laboratories. The prohibitive cost of software packages limits many schools to utility packages incorporating word processing, databases, spreadsheets and graphics. Specialised packages for individual subjects are rarely available within the school.

Although teachers have been given opportunities to develop their skills, many are still not comfortable with computers, particularly as many of their students may know more than they do. New initiatives in teaching require time and enthusiasm to develop. How many teachers in any country have sufficient time to devote to becoming familiar with computer software and developing applications? Some will make time, most will not.

It was the view of this regional co-ordinator that, too frequently, a school's "computer guru" hinders rather than helps his or her colleagues' acquisition of computing skills, for example, by not passing on information about courses which are being offered and which would be relevant. In many schools, informatics in the classroom is restricted to learning introductory keyboarding skills, word-processing, and an introduction to selected general applications, while mathematics and computer science classes retain traditional computer science with a strongly mathematical orientation.

Many of the views expressed by this typical co-ordinator were echoed by participants at the educational informatics conferences. Most participating teachers, irrespective of whether they are from the United States of America or Australia, express their frustrations which are created as a result of sitting through exciting multi-media presentations of specific curricula components, knowing that these packages are beyond the financial resources of their individual schools. Many comment that their schools contain too high a proportion of teachers with minimal interest in informatics and little recognition of the need for their students to become involved with this developing technology.

Attitudes of schoolgirls towards informatics

Given that most school have at least one computer laboratory, how successful are school computing classes? We attempted to answer this question by asking schoolgirls. In a qualitative study of secondary girls' attitudes towards informatics, thirty-two girls, aged 13 to 17 years, were interviewed. Girls from four schools, encompassing all combinations of public, private, single-sex and co-educational schools were included. The girls to be interviewed were selected by the schools; the contact teacher was asked to identify girls who had the ability to study informatics at higher levels, but who had not necessarily displayed an interest in the topic.

The girls were asked about their career aspirations, whether they enjoyed their school informatics studies, whether they used a computer at home, and what they thought about careers in informatics. The motivation for the study arose from an earlier series of interviews which indicated that the perceptions of computers and computing held by university computer science and informatics students were markedly different from those of women working in the field [9].

When asked about their informatics studies at school, the schoolgirls typically described their classes, and informatics in general, as 'boring', and not an area to study voluntarily. Their views apparently related to their experiences in learning keyboarding and basic computing skills in a formal classroom situation. In none of these four schools had teachers managed to make informatics classes enjoyable and motivating for the majority of female students. It appeared that students were taught using standard class exercises. While this approach is used in many subjects, it is clearly turning girls away from formal informatics studies. The only three students who mentioned specifically that they had enjoyed their informatics studies were older students who reported using a variety of application packages in the later years of their schooling.

When asked about their home use of computers the girls who had access to a computer at home become enthusiastic, describing the advantages of using a computer to complete assignments. Word processing was the main application of home computers, however, graphics packages, mathematical work and spreadsheets were being used.

The girls interviewed appeared not to relate what they did at home to the formal informatics instruction of the classroom, and did not consider their home computer use to be informatics.

The problem appears to be one of relevance. These girls enjoy using computers at home to achieve their own objectives, but may not enjoy formal tuition in informatics when the class objectives are not consistent with their own immediate goals. They do not see any similarity between what they do at school and what they do at home.

In contrast to the bleak picture conjured up by the interviews, one local high school has its computer laboratories filled with girls at lunch time. The girls, and a few boys, use the computers to prepare high quality assignment submissions, where as, a decade ago, it was game playing boys who filled the room at lunch time. Enthusiastic schoolgirl users such as these can be found in schools world-wide where computers can be used to enhance the quality and presentation of projects which girls find interesting and relevant [10].

The schoolgirls we interviewed did not see informatics as an attractive career option, believing it to involve working in isolation and sitting at a screen all day, although they admit to having little knowledge of the careers available. In contrast, women using informatics in the workplace describe their work as challenging, varied and requiring constant interaction with people. In fact, the women working in informatics identify the very same job characteristics which the schoolgirls were seeking in an ideal job. It seems to be lack of knowledge about the real world of informatics, rather than the true nature of informatics which is turning girls away.

Victorian certificate of education

A few years ago the Victorian government introduced the new Victorian Certificate of Education (VCE) which consists of 24 units of study completed over the final two years of secondary education. The units completed in the second year of the VCE are assessed using Common Assessment Tasks (CATs), of which approximately one-third are examinations with the remaining tasks being essays or projects completed over a number of weeks. Students are required to submit plans and drafts of their work prior to the final submission. The intention is to encourage the development of skills which are needed in the workplace. Most students have discovered that computers are fantastic tools for completing these tasks as they can present revised drafts without re-writing. Additionally, computers have spelling checkers and word counters, and they can be used to produce graphs and diagrams as well as to analyse data. Some students have their own private computers while others rely on the school computing facilities. Although this was not a stated intention of the structure of the VCE, the increased recognition of the role of computers has certainly resulted from the changed curriculum and assessment demands. Students have discovered that computers are very useful tools that enable them to produce better quality work with considerably less effort than is possible without these facilities. Even without the VCE, schools could learn from this experience and devise more assessment tasks which would benefit from the development of skills in informatics.

The importance of diverse experiences
While it is not possible for all students to have their own personal computer, it is possible for teachers to maximise the limited access time available by ensuring that their students have a broad range of informatics experiences. The importance of diversity was demonstrated in a recent study of approximately 200 secondary girls drawn from both single-sex and co-educational schools. Jones and Clarke [11] found that the girls in the single-sex schools had more positive attitudes towards computers than did the girls in the co-educational schools, such differences being independent of whether the schools were government or non-government schools. However, a more detailed analysis of the data showed that the girls' attitudes were influenced by the diversity of their experience with computers, rather than, as often implied in previous research, the sheer amount of computing experience. It seems that girls in single-sex classes were mastering a particular computing application, then approaching the teacher to be allowed to use a different application. In co-educational classes girls were mastering an application and continuing to use it, as they felt comfortable with that application and were reluctant to compete with the boys for either the teacher's attention or the opportunity to use a different application. The data clearly indicate the importance of using time in informatics classes to provide students, both girls and boys, with experience with a wide range of applications to enable them to develop an appreciation of the uses of computers and more positive attitudes towards using them.

CONCLUSION AND RECOMMENDATIONS

Whilst schools have been successful in providing students with opportunities to use computers, they have been less successful in providing those opportunities across the curriculum, and apparently have failed to stimulate interest in the exciting world of informatics, particularly among female students. This could be addressed in the following ways.

Resources. The resource issue needs to be addressed, whether by the provision of additional resources or the development of shared computing laboratories with shared software. Increased sharing is possible. At the upper secondary level in Victoria groups of secondary schools are re-organising into middle school campuses teaching years 7 to 10 and secondary colleges campuses teaching years 11 and 12. By sharing facilities across a large multi-campus organisation rather than working as separate schools, there is a greater potential for sharing hardware and software.

Class time. In the limited time available to teachers using computing laboratories, it is important to provide students with a variety of different informatics experiences. Rather than focussing on one application and thoroughly mastering its use, teachers may be better advised to allow students to have an introduction to a variety of different computer applications.

Relevance. Girls in particular need to see the relevance of learning to use a particular application. By introducing a range of applications through the development of a group project, students can not only see the need to develop skills in informatics, but can also learn to work as members of a team, developing the interpersonal skills so essential to most jobs, and especially to jobs in informatics.

Not only informatics. The study of informatics should not be limited to informatics classes. Students should been encouraged to use their informatics skills to collect, analyse, organise and present information across all areas of the curriculum.

Image. The common, but inaccurate image of careers in informatics as boring and socially isolating needs to be replaced with a more accurate image indicating the extent to which these workers enjoy the challenge, variety, problem-solving, people-oriented nature of their jobs.

Girls' informal use of computers suggests that informatics is again becoming gender neutral, and that schools must consider means of stimulating interest in formal informatics classes.

REFERENCES

1. Clarke, V.A. (1992) *Strategies for involving girls in computer science.* in In Search of Gender Free Paradigms for Computer Science Education, C.D. Martin and E. Murchie-Beyma, (eds.). NECC Monograph: Eugene, Oregan. 71-86.

2. Clarke, V.A. and S.M. Chambers. (1989) *Gender-based factors in computing enrolments and achievements: Evidence from a study of tertiary students.* Journal of Educational Computing Research. **5**, 409-429.

3. Culley, L. (1988) *Girls, boys and computers.* Educational Studies, **14** (1), 3-8.

4. Jones, S. (1991) *Why do we want to see more women in computing?*, in Women into Computing, G. Lovegrove and B. Segal, Editors. Springer-Verlag: London. 59-65.

5. Eagly, A.H. (1987) *Sex differences in social behavior: A social-role interpretation.* Hillsdale, NJ: Lawrence Williams Associates.

6. Clarke, V.A. (1990) *Sex differences in computing participation: concerns, extent, reasons and strategies.* Australian Journal of Education, **34** (1) 54-68.

7. Lovegrove, G. and Hall, W. (1991) *Where are the girls now?* in Women into Computing, Lovegrove, G. and Segal, B. (eds.) Springer-Verlag: London. 33-45.

8. U.S. Department of Education. (1993) Office of Educational Research and Improvement, National Center for Educational Statistics, Washington, D.C.

9. Teague, G.J. and Clarke, V.A. (1991) *Fiction and fact: Students' and professionals' perceptions of women in computer science,* in Women, Work and Computerization: Understanding and Overcoming Bias in Work and Education, I.V. Ericksson, B.A. Kitchenham, and K.G. Tijdens, (eds.). North-Holland: Amsterdam. 363-375.

10. Yewlett, H. (1992) *Everyone does information technology in Ystalyfera!* in: the proceedings of Women into Computing. Teaching Computing: Content and Methods. Keele University, Staffordshire: p. 215-221.

11. Jones, T. and Clarke, V.A. (1993) *Girls and computers: An assessment of the advantage of single-sex settings.* Presentation at Networking in the 90's. The 2nd Women in Computing Conference, Melbourne.

Joy Teague completed her high school education in 1964 and started work in a bank. Eight years later, after completing a part-time B Sc, embarked on an academic career at the Gordon Institute of Technology, and, in 1977, transferred to Deakin University. Currently a Senior Lecturer with the School of Computing and Mathematics at Deakin, she is enrolled in her fourth part-time degree programme, studying for a PhD. Primary research interest is gender related perceptions of computing careers.

Val Clarke completed her secondary education at a private girls' school before majoring in psychology in Arts at Melbourne University where she was awarded a BA (Hons) and an MA. She lectured in Humanities at the Gordon Institute before leaving to become a full-time mother on the birth of the first of her two daughters. She returned to work at Deakin University in 1979 as a Senior Tutor. After many years of teaching and research and completing a PhD at Deakin, she is now a Senior Lecturer. Val's major research area is gender issues, with a special emphasis on the increased participation of women in mathematics, science and technology.

22

Learners and their expectations for an integrated informatics environment

Alison I. Griffith
Gene Diaz

University of New Orleans, USA

Cassandra Murphy

Tulane University, USA

ABSTRACT

This paper explores students' expectations for an integrated informatics environment. We interviewed students from four secondary schools in New Orleans, LA. Our data show an unsettling lack of critical investigation in courses about informatics technology. Nonetheless, students have developed critical understandings of computers in society based on futurist portrayals of society on television and in popular films. Their expectations are mediated by social class. Working class and minority students are more likely to see the social contradictions of informatics technology than are other students.

Keywords: classroom practice, curriculum policies, knowledge representation, research, visions

INTRODUCTION

Issues of equity and access, shaped by the inequalities of gender, race and class, have been identified by many educators as central to understanding the uneven development of informatics implementation in U.S. schools. We suggest there is an additional and perhaps primary issue. In schooling, courses and resources are focused on acquisition of the technical information which precludes the development of an informed and critical exploration of the impact of informatics on society [1, 2, 3].

This pilot study explores students' expectations for the shape and character of informatics in their future. As we have explored this topic with secondary school students in New Orleans, Louisiana, we have discovered an unsettling gap in the computer curriculum. The social organization of computer knowledge is rarely explored in secondary school courses. Students have little basis in their schooling for developing an informed and critical stance towards the impact of informatics on society. In our research, we found that secondary-school students had developed a vision of an informatics-based future derived from futurist portrayals on television and in film. Students' describe both positive and negative consequences of future informatics uses in society. Our research suggests that technical skills-oriented curricula displaces a critical and informed discussion within schooling about informatics - a dialogue which is currently conducted between the student and the media. This dialogue is mediated by social class positioning - working class and minority students are more likely to see the social contradictions of informatics than are other students.

EQUITY, ACCESS AND SOCIAL CLASS IN EDUCATION

As we move toward the 21st Century, schools in the U.S. continue their uneven training and education of students for a society in which informatics plays an ever-increasing role. Consistently, researchers have shown that issues of equity and access in educational computing are directly related to students' ability classification, gender, race/ethnicity and socio-economic status [4, 5, 6, 7]. Becker [4] suggests that issues of student access are predicated on the availability of computers and, as importantly, on teachers' ability to incorporate computer-oriented analytic thinking and problem solving into their classroom activities. Kirby and

Styron [8] examine the issue of access to technology as a function of ability, race and gender. In Louisiana, computer access appears to be mediated by academic expectations held by teachers and administrators. We can expect, therefore, that student knowledge of and expectations for the limits and possibilities of informatics will reflect the differences of equity and access that permeate educational technology courses in the U.S.

The equity and access literature is limited in its ability to contextualize educational computing within the larger institution of schooling. This research gives us some understanding regarding school-generated differences in student knowledge about educational computing and yet, describes only part of the picture. These studies rely on a liberal conception of schooling with a focus on individual rights and responsibilities. Within this framework, problems of equity and access are conceived of as technical problems of student access and teacher training.

When we place the concerns of equity and access in the context of schooling, we find they do not stand in isolation from, but rather mirror the relations of inequality constructed within education [9]. Social class membership is strongly constitutive of school outcomes [10]. The social relations of class position working class and minority students differently from middle class students. Most working class students have less investment in the hegemonic structures of schooling and their marginality provides for the possibility of a critique of schooling. This is not to claim that the critique is necessarily a sophisticated one. Rather, it is to suggest that the oppositional stance of working class and minority students provides the basis for a critical stance toward schooling. In the case of educational computing, our data show that working class and minority students are more likely to see the social contradictions inherent in informatics than are their middle class counterparts.

Bowers [1] claims that the emphasis in educational computing on technology and individualism inhibits critical analyses of informatics technology. Computing curricula that do not raise questions about the limitations of knowledge available technologically encourage "digitized thinking" and preclude critical discussions of the complexities of informatics in society. As described by Cummins [11, p.viii], the computer is conceived of as an "adjunct transmitter of knowledge and skills".

The students we interviewed were critical of technological processes taking place in society. As we will see below, the critique is based first in the social relations of class and race that marginalize working class and minority students, their families and their communities; and second, in the futurist portrayals of television and media which both glorify technological possibilities and introduce the expectation of social distress. Critical knowledge about informatics technology is only minimally grounded in students' schooling experience.

METHODOLOGY

We conducted group interviews with students in four secondary schools in New Orleans, LA USA. The schools included two public magnet schools (known here as Mondrian and McClintock Schools), a private school (Pious School) and a comprehensive public secondary school (Downtown School). Mondrian and Pious Schools enroll primarily
students from middle income families; McClintock School draws from a wide range of families but includes a substantial number of students from low income families; Downtown School students are typically from families with low incomes. All the students we interviewed at the public schools were African-American while those at the private school were European-American. The students ranged in age from 14 to 19 years (Grades 9, 11, and 12) and were enrolled in computer literacy or computer science courses. We also informally interviewed their teachers.

The group interviews were conducted during class time and typically held in the school library. One interview was held in one corner of the classroom. The University of New Orleans Human Subjects Review regulations required we interview only students whose parents signed consent forms. In total, 30 students were interviewed.

The interviews were tape recorded and selectively transcribed by the researchers. We looked for themes and patterns within the interview data that were consistent across all the student interviews. As well, we reviewed the literature in the area of educational computing in order to discover areas of overlap with our data.

The student population in New Orleans should not be seen as representative of student populations in the rest of the United States. First, the diversity of the U.S. population means that there are vastly different ethnic, racial and class configurations across the country. Second, Orleans Parish is an urban centre which has an African-American majority and a wide range of minorities including Hispanics, Asians and European-Americans. Third, the different race/ethnic groupings, such as the African-Americans, are found throughout the class structure of New Orleans. Fourth, the graduation rate for Louisiana is one of the lowest in the U.S. The students we interviewed are more likely to be those who will complete their secondary school education.

This has implications for the relationship between our findings and other research in the area. For example, our research will not show the wide range of differences in educational computing by race or gender. However, class differences are very evident. Thus, we have focussed on social class as it shapes the different experiences and expectations of students about informatics.

DISCUSSION

The interviews with the students brought into view several patterns of use and understanding. Computer access and use patterns supported the findings of the literature on equity and access cited above. Other patterns included differences in students' expectations for the role of informatics in their future and the role of the media in constructing these expectations.

Differences in access

Our interviews showed patterns of educational computing shaped by social class. Differences in access and resources available to the four groups of students were closely linked to family and school resources - social class. At Mondrian School, all of the students we interviewed had computers at home. So, too, did the students at Pious School. However, only one of the students interviewed at McClintock School and one student at Downtown School had a home computer although several students claimed to have access to computers through friends.

The amount of time students spent on computers reflected social class resources available to them through their schools. At Mondrian School and Pious School, the students worked at computers for 30 minutes, three or four days of the week. McClintock School which focuses on mathematics and science had just cut the amount of time in which students are engaged in computer courses to once per week (20% of the students' mathematics grade). Downtown School students spent the least amount of time on computers and had fewer computer resources to draw on than did other students with whom we spoke. They spent little time actually using computers because most of the classroom computers were not functioning. Often they had to share the computer with several students. During this school year, they have had five substitute teachers in their computing course. Their current teacher was uncertified in the subject area but was concerned that students be able to finish the computer science course.

Differences in critical knowledge

Obviously, the different school experiences and the range of resources available to them will have an impact on these students' knowledge about computers. What is not as obvious is the relation between their critical understandings about computing and their social class.

We asked students to tell us about a future in which informatics was integrated into their everyday lives. They described a number of scenarios they thought might be possible. One of the scenarios included schooling in the future. The students said that teachers would become obsolete and that all learning would be done through computers. The students at Pious School were concerned that this would reduce the experiential learning that is part of the school experience. Pious and Mondrian students said there would be a computer on every desk and that every student would have a laptop and a computer at home. Generally, students at Mondrian and Pious described technological changes based on informatics as "...for the best". On the other hand, students at McClintock and Downtown were much more pessimistic about an informatics-based future. Downtown and McClintock students envisioned a future without schools where students could educate themselves at home through computer-assisted instruction. The McClintock students described a scenario in which they did their lessons on the computer during the day and then went out with their friends in the evening.

For our purposes here, it is the underlying conception of computers as knowledge transmitters that is of interest. This conception of computers supports the "transmission-internalization duet" [11] p viii, of current curricular models. While most pronounced at McClintock and Downtown, this vision of informatics was held by all of the students. Noble claims that computer-based education research and development is grounded in a conception of "sophisticated man-machine systems" [3] p 188, - a conception which shapes these student expectations of informatics as well. The computer is seen as a machine that (potentially) contains all the school knowledge students need. In the future described by the McClintock and Downtown students, the computer-student link will separate students from schools but not from their communities.

Computers and the future

We asked the students two kinds of questions about their expectations of future uses of informatics. First, we asked them to describe computers in the 21st Century. We reminded them that 1994 was six years from the turn of the century, and asked them which of their visions of a future including informatics would be part of everyday life.

The student descriptions of informatics futures were revealing. They described a future that included talking books, computers on each student desk, computers that would recognize householders' voices for setting home security systems; voice recognition computers that produced hamburgers on demand; computer-controlled flying cars; medical advances and so on. Their descriptions of the year 2000 were more technologically moderate and included voice-recognition computer systems, increased police surveillance strategies, more sophisticated software for training pilots, and communications systems that included video cameras.

We saw no basis for their critical visions either in course materials, or in curriculum guides [12], all of which are skills-based, some of which teach a technological history of informatics.

The wide-ranging student descriptions of informatics technology for the future ranged from the fantastic to the concrete. The student descriptions were replete with excerpts from television and films that represented the future. For example, several students remarked on the American Telephone and Telegraph (AT&T) commercial showing a mother checking on her sleeping baby using a video monitor. Other students mentioned films such as *Total Recall* as an example of the future of informatics technology. Indeed, man-machine films and television programs such as *Star Trek: The Next Generation* and *The Jetsons* were cited as the basis for their conceptions of an informatics future.

When asked if they could see any drawbacks to the technology they had envisioned, the responses varied by school and social class. For example, most of the Pious students thought that informatics technology would be congruent with their careers and interests. The Mondrian and McClintock students asserted that technological advances would mean people would become lazy. The Mondrian, Downtown and McClintock students suggested it would make work easier. They also described other, more formidable consequences for informatics technology. One student at Pious and several students at Downtown and McClintock suggested that computers would contribute to the development of advanced war machines. The McClintock students and the Downtown students had the most pessimistic expectations for informatics. Downtown students suggested that people would be "laid off" because of robotics. Another suggested that the future was jeopardized by "mad scientists" and the possibility of nuclear warfare. These doomsday scenarios indicate a mistrust of both technology and the technologically adept.

With some exceptions, our data point to some intriguing differences in critical reflection about computers in society. First, the middle class students tended to describe computers in neutral ways. The working class students, on the other hand, described informatics as having both positive and negative consequences. Second, the students from lower income families were more focused on the social impact of computers in their lives while the middle income students described computers as tools for their future careers.

CONCLUSION

We suggest that the differences are linked to the different social positioning of the two groups of students. Low income students (the majority of those we interviewed from McClintock and Downtown schools) are more 'at risk' of failure in the schooling process. In New Orleans, working class neighborhoods are often unsafe and, at times, public schools are visited by street violence. One student at McClintock said: "Nobody won't be here if peoples keep killing each other. Nobody going to be here to see no computers." As well, the relevance of the school curriculum to the lives of inner city students is questionable. There are often significant differences between the school curriculum and their lives outside school. Thus, points of contradiction are present in their schooling experience - contradictions that generate oppositional and critical possibilities. In contrast, the middle income students' experience of schooling (at Mondrian and Pious schools) is more congruent with, and linked to individualist understandings of their lives outside schooling.

Of equal importance in developing particular visions of an informatics future are differences in television and film viewing patterns between classes. The Downtown and McClintock students mentioned media-based scenarios more often than did the Mondrian and Pious students. As the Louisiana secondary school curriculum for educational computing focuses on skill-building and ignores critical analysis, students projections for their future in informatics technology develops through their exposure to the media. Their expectations for an integrated informatics environment are shaped outside of schooling, in television and films in which computers are transmitters of objective, factual knowledge. In both settings, the possibilities for a questioning stance about the impact of informatics on society are truncated or distorted.

Our research shows that students have a critical understanding of informatics that is mediated by the social relations of class. A critically-oriented curriculum for informatics technology courses has been recommended by theorists such as Bowers [1] and developed by educational researchers such as DeVillar and Faltis [6]. The implementation of such curricula in US schools would provide students' with an opportunity for critical reflection on technology.

REFERENCES

1. Bowers, C. A. (1988) *The cultural dimensions of educational computing: Understanding the non-neutrality of technology.* New York: Teachers College Press.

2. Griffith, A. I. (1992) *Educational computing in the elementary classroom.* Education and Computing. **8**, 53-60.

3. Noble, Douglas D. (1991) *The Classroom Arsenal: Military research, information technology and public education.* New York: Falmer Press.

4. Becker, H. (1990) *Computer Use in United States Schools: 1989 - an initial report of U.S. participation in the I.E.A. computers in education survey.* M.D. Center for Social Organization of Schools. Baltimore MD: The Johns Hopkins.

5. Cusick, T. (1986) *Beyond the Star Trek Syndrome to an Egalitarian Future: Where No One has gone before.* Washington, DC: NOW Legal Defense and Educational Fund. ERIC Document Reproduction No. ED 304 130.

6. Devillar, R. A., and Faltis, C. J. (1991) *Computers and Cultural Diversity: Restructuring for School Success.* Albany NY: State University of New York Press.

7. Malcolm, S. M. (1988) *Technology in 2020: Educating a diverse population* in Technology in Education: Looking toward 2020, Eds R. S. Nickerson and P. P. Zodhiates, 216-30. New Jersey: Lawrence Erlbaum Associates.

8. Kirby, P. C., and Styron, R. (1993) *Access to technology as a function of ability, race, and sex.* Paper Presented at the Annual Meeting of AERA. Atlanta.

9. Apple, M. (1988) *Teachers and Texts: A political economy of class and gender relations in education.* London: Routledge and Kegan Paul.

10. Griffith, A. I., & Smith, D. E. (1990) *'What did you do in school today?' Mothering, schooling and social class.* Perspectives on Social Problems **2**, 3-24.

11. Cummins, J. (1991) *Foreword.* In Computers and Cultural Diversity: Restructuring for school success, R. A. DeVillar, and C. J. Faltis, vii-ix. New York: SUNY.

12. Louisiana State Board of Elementary and Secondary Education (1985) *Computer Science curriculum guide.* Baton Rouge, LA: State of Louisiana.

Alison Griffiths is an Associate Professor in the department of Educational Leadership, Counselling and Foundations at the University of New Orleans, Louisiana, USA. She was trained in Canada at the Ontario Institute for Studies in Education as an educational sociologist. Currently she teaches qualitative research methods and social foundations of education courses. Her research on informatics in education explores the social and critical features of educational policy and practice and investigates the social relations of discourse that construct informatics in education.

23

Societal and organizational influences on integration: what about networking?

Betty Collis

Faculty of Educational Science and Technology
University of Twente, The Netherlands

ABSTRACT

This paper reflects on the thesis that social and organizational forces have a great influence on the degree to which and the way in which the student encounters communication and information technologies in the secondary school. To expand on this thesis, a selection of social impulses that have served or are serving as a technology push in schools are considered, from "computer literacy" to "Internet literacy". As a counter force, some key aspects of the ways schools as organizations shape and constrain access to technology are reviewed. The movement toward "community networking" and "virtual networks" is considered as a current example of how complex social developments from outside education can influence the ways young persons and teachers experience communication and information technologies in schools. The likelihood of information technology becoming deep rooted enough to significantly affect educational practice is considered.

Keywords: networks, communication, future developments, innovation, social issues

INTRODUCTION: SCHOOLS IN THE YEAR 2000

Visions and agendas for "schools in the year 2000" are beginning to appear in a number of countries. A conference in April 1994, for example, in the United States brought together nearly a thousand educational decision makers to discuss such an agenda. The assumption that was continually expressed throughout the meeting was one of radical change in education as it is now organized, moving from "buildings, lecturing, reading, and testing" toward the model of a "global village school":

> "...a round-the-clock community of learners not necessarily bound by a place or building ... where learning occurs where it is appropriate: not only at its headquarters but also in the homes of its students, in the activities of its teams and clubs throughout a region, and in the far-flung sites that the student may reach by way of field trips and electronic connections". [1]

According to the speakers at this conference, the standard by which a student is judged to be well educated will be the extent to which he or she can effectively use the tools of the information age. Critical to such a vision, driving it as well as serving it, is the powerful metaphor of an "information superhighway".

To what extent is this vision being driven by forces - social, political, economic, philosophical - outside the professional education community? To what extent is the way in which the secondary-school student has the opportunity to make use of communication and information technologies influenced by such forces? It is part of the thesis of this paper that such "outside forces" exert a major influence on the educational deployment of communication and information technologies. This influence however is constrained by another powerful set of influences, the organization of secondary schooling itself, which also exerts a powerful effect (often a counter effect of resistance or inertia) on what happens with technology and the learner. This paper also briefly discusses some of these institutional influences.

In these two contexts, "networking" is seen as a powerful idea in society which is now gaining new impetus in terms of its potential influence on education and on the ways that communication and information technologies are applied in education. Is networking another example of an externally pushed idea shaping educational agendas, eventually to be neutralized by the resistance of the school as an organization? Or might networking emerge as being powerful enough educationally to bring a lasting change in the use of communication and information technologies in schools?

SOCIETAL INFLUENCES ON COMMUNICATION AND INFORMATION TECHNOLOGIES IN SCHOOLS

It can be argued that many major influences on the motivations for communication and information technologies in schools have not come from the educational community but from outside it. Five of these powerful ideas, two already losing momentum and three gaining power, are "computer literacy", "back to the basics", "equity and mobility", "cyberspace and the information superhighway", and "just-in-time learning".

Computer literacy
A number of analyses [see, for example, 2] have been done of the impetus for computers in schools and the phenomenon of "computer literacy", which moved at incredible speed throughout the world in the early 1980s and triggered mass-scale acquisition of computer hardware, software production, teacher training, new curriculum, and other manifestations. Generally these analyses show that calls for "computer literacy" were stimulated strongly by arguments such as "every child must be computer literate to be employable" and "schools must make sure students are computer literate in order for the larger society to maintain economic competitiveness".

Typical of the response to such societal pushes was for a country or region to put in place a large-scale effort to get computers into schools and to stimulate some initial training and software provision to accompany the arrival of the computers in the school. However, planning and provision for long-term support, on-going critical evaluation, and for the follow-up needed to bring teachers to the point of actually integrating computers as an established part of their teaching was often a second-layer of consideration, not part of the range of analysis of the initial computer-literacy advocates. The results can be seen in the frustratingly low levels of use of computers in secondary schools outside of courses that teach about computers [3]. But the social push of "computer literacy" was strong enough to set up an infrastructure in schools for computer use, a legacy which remains even as the external calls for being "computer literate" no longer much appear.

Back to the basics

Another factor which has made a strong impact on education in general and on the use of information technology in particular is the reform-movement wave which regularly arises in different societies. One example in particular is the "back-to-the-basics" movement which was particularly influential in the United States but has had manifestations under different names in other cultures as well. "Back to the basics" was a reaction against "what was perceived as a growing liberalism and permissiveness among educators" while "wearing the mask of national interest and competitiveness" [4]. The driving agent of change was not educational analysis, but legislated reform stimulated by special interests, politics and economics [5]. The general assumption was that teachers were the problem; information technology was often seen, explicitly or implicitly, as a way to increase the efficiency of education, by providing better management and better designed instruction than the teacher was delivering with his or her resources. Although "back to the basics" has faded as a motivating buzzword in American circles, its influence still implicitly remains in terms of "integrated learning systems" and CMI (computer-managed instruction), and its social and organizational manifestations are still occurring in many countries, under movements called by names such as "accountability" and by mechanisms such as the publication of "league tables" that rank order schools on the basis of performance on standardized tests.

Equity and mobility

In contrast to the above two impulses, a theme which is currently very strong in Europe as well as the United States is that of providing equal opportunity for education (and later, for employment) regardless of the person's geographic location or personal circumstances. In Europe, this theme motivates many major initiatives; for example, the evolution of "flexible and open learning" and the idea of a "trans-European network for education and training" are explicitly based upon the policy goal of equal access throughout Europe to education and training [6]. In the United States the "vision" stimulating the emerging policy of the President's Executive Office of Science and Technology is stated as "To ensure every American access to excellence in education and training regardless of location, background, or personal situation" [7]. Both of these impulses have large political and social motivations and agendas. They also are explicitly based on the assumption that such equity and mobility will result in a stronger workforce and a more productive economy. Few of the prominent policy shapers appear to be from educational backgrounds.

A direct consequence of these equal-access impulses is the expanding interest in more flexible learning opportunities, including a heightened interest in distance education, in that flexibility of time and place become necessary if equity in opportunity is to be developed. Communications technologies in particular are a major strategic key in providing flexible access to quality education, and as such are central now to a great number of initiatives in Europe, North America, Australia, and throughout the world. Institutions with well-developed distance education systems are now being joined by traditional educational institutions wishing to diversify and broaden their offerings, with the link between them being the use of communications technologies.

Cyberspace and the information superhighway
Another impulse which in the last year has burst forth in the popular media is the idea of cyberspace and the information highway. Although these terms are metaphors, they are powerful ones and suddenly are being taken seriously by society around the world. A large reason for this is the popular discovery of the Internet system, a network of networks sharing a common protocol which has been in existence for nearly 30 years but within the last one or two years has suddenly become a "household word". The number of networks in the system has doubled between 1992 and 1993, a new network is said to be connecting every ten minutes, and perhaps 20 million people now have access to the system. A sense of exhilaration stimulates many of those discovering this "cyberspace", a sense of "the freedom of the open road" to cite one of the many cliches now appearing. Not being able to be freely and equally connected to the Internet system or other networks is now being seen as a denial of a "basic right" [8]; not having access on demand is seen as being cut off from the information age, from the "two-fold emancipation ... from the limited nature of exchanges with his or her (immediate) fellows, and ... from the multiple servitude of distance in time and space" [9]. Like computer literacy a decade ago a new goal for education is being advocated (see also the definition of what being "well educated" will mean in the schools of the year 2000 cited at the start of this paper): "not knowing how to use the Internet will be as grave a deficiency as not knowing how to read" [10].

This statement was made by a businessman, not an educator. But clearly, the "superhighway" impulse is shaping new applications of communication and information technologies in schools.

"Just-in-time" learning

Another idea is becoming more and more prominent in corporate settings and in large-scale training enterprises: the idea of "just-in-time" learning, where the learner, usually at his or her workplace and even while carrying out his or her ordinary work, has access to learning support when it is perceived to be useful. This impulse is driven by financial motivations to a great extent, in that it is assumed such workplace-based training will reduce costs associated with "going away" to traditional training courses, and also that it will result in more efficient training given the proximity of need, training, and opportunity for transfer.

This impulse is already suggesting changes for secondary education. The just-in-time approach implies that more and more responsibility comes to the individual in deciding when a learning episode is necessary and what resources are most appropriate for it, and that the responsibility of the trainer and employer will shift to that of provider of a range of learning resources and opportunities [11]. This approach also by definition calls for sophisticated resources and the use of communications technologies. Secondary school education is being called upon to prepare students for this type of future; to "learn the survival skills that will keep them employable through life" which will be heavily based on, in a self-directed way, "learning how to learn in the electronic workplace" [11].

COUNTER-BALANCE: THE SCHOOL AND ORGANIZATIONAL CONSTRAINTS

Societal impulses such as the five noted above do have a strong influence on what happens in the school with respect to communications and information technologies. But not as much as they might. The school organization serves as a solid resisting wall to rapid change impulses, a phenomenon which has been well documented in many countries and contexts. Space restraints only allow a summary of points here.

Organizational constraints in the school system

One cluster of organizational factors that are constraining the deployment of the current impulses related to equity and mobility, learning in cyberspace, and just-in-time learning are those related to the organizational identity of the school as an institution, issues related to boundary articulation and domain determination [12]. Institutions have their own accountability and cultures and perimeters; "bringing the world into the school" or other similar paradigm shifts, no matter how compelling, do not fit with existing organizations and their ways of defining themselves if they imply other corridors of authority and decision making. And this is not a matter of "fault" or limited vision, but rather a general survival trait of benefiting from consistency and established roles and structures. Secondary schools not only reflect the general resistance of an institution to change (and perhaps to lose some of its control over itself) but also are limited by other real and on-going expectations of society that continue even as impulses come and go - that students will learn a particular curriculum, will pass their tests, will earn their diplomas, and will make an effective transition, in as efficient a way as possible into the workplace.

These expectations can come in direct conflict with the attempted execution of social impulses such as the five described above. We have seen this in our decade of experience with computers in schools: except for courses where using computers becomes part of the testable curriculum, computer use remains an option, an add-on, an experience most likely to be available to those fortunate enough to have a teacher motivated to extend or enrich traditional instruction with computer use. Teachers are too busy with the demands of the curriculum, of the upcoming examination, of the challenges involved with guidance and management of their students, to handle the changes needed to reorganize their teaching in order to exploit the potential benefits of technology.

A new organizational constraint: the computer establishment in the school

The above sorts of constraints within the school and school system are well known; a more recent constraint comes from within the computer-use sector in the school itself. In the decade that computers have been in secondary schools, most schools now have a well-defined computer course and teachers associated with those courses. The computer courses, called by many different names in different countries, require extensive use of the school computer laboratory, and the school computer coordinator now has a well-thought through system for managing the computer laboratory, its local-area network, its security, and its maintenance.

In this environment, the teacher who wishes to explore some application of computer use with his or her students in a subject area such as language or science has difficulty. The laboratory is booked and busy and may be in a location far-removed from subject-area classroom; how can some of the students use the computers as tools during their writing or science activities? How and when can the teacher explore new uses of the technology? Not while the students are present, surely; but this means late after school, and then it may be that the coordinator wishes to lock up.

These collisions are already well know with respect to the attempted integration of computers into traditional subject areas. Recently we have seen them take on a new form in the context of teachers wishing to make use of communications technologies in schools [13]. It is even more problematical to try to connect students to an external network service or system than it is to try to find a way for them to use the computer lab for a familiar task such as word processing. The computer coordinators are specialists now in information technologies and Local Area Networks (LANs), but may have no experience with wide-area networking and its tools and protocols. The school administrator has learned how to gauge costs of computers, but has no idea how to anticipate the costs of students and teachers "navigating the Information Superhighway".

We encountered these problems in considerable magnitude when we carried out a project in 1993 that involved stimulating communications technology use in various vocational-education schools [13]. Despite the fact that we offered the participating teachers two computers and modems each, one for school use and one for home use, the teachers ran into serious problems trying to make use of these computers for telecommunications uses. The insurance arrangements that some of the schools had required that all computers be in the computer laboratory; thus the teacher could not use his or her computer in the classroom or work with it there after school. In other schools, the computer coordinator did not want to deal with the maintenance of any computer system outside the one which had been chosen for the school; thus we had to purchase new systems, different for each school, to match the idiosyncrasies of their school configurations. Wide-area connections interfered with the LAN software in some of the schools, and disengaging a computer from the LAN to separate telecommunications use was also not well accepted. Schools did not have a way to separate telephone charges for telecommunications uses from other telephone charges; or budget procedures to approve a project when the cost of the project was not specified in detail in advance (difficult to do with on-line use: the more successful a project, the more use will occur); or procedures to reimburse teachers for the on-line charges they incurred at home in their on-line lesson preparation.

In time, these sorts of organizational barriers will sort themselves out into procedures such as now are in place for the mainstream computer laboratories in schools; but then the next opportunity made possible by technology (perhaps more use of video and audio-graphic conferencing?) will encounter a similar wave of organizational frustrations.

NETWORKING: THE NEWEST SOCIETAL IMPULSE AND ITS CHANCES IN SECONDARY EDUCATION

Community networking

A new impulse is developing momentum in society: the concept of networking. Networking is an interesting word, in that it is both a noun and a verb, and can be used to refer to human associations, computer associations, or both. Within its technical associations it is used to refer to a number of different types of systems. It is also a powerful metaphor, one becoming more and more compelling given the rapid rise in "virtual communities" made possible by communications technology. These range from the countless local or regional bulletin board networks to the discussion groups of strangers that somehow evolve a sense of community through on-line networks. Many times these on-line groups are loosely defined, with no articulation of membership criteria or even no real sense of who is moving in or out of the community. But the communities that are having a more profound effect are those in which the on-line communities are a defined group who enrich and extend themselves through their on-line interactions. These sorts of groups are proliferating among teachers.

The LABNET Project based at TERC in Boston is a prime example [14]. Over a two year period, over 200 secondary-school physics teachers made use of computer networking to strengthen their human networking relative to their communal work on curriculum development in physics, and their participation in various classroom-based projects in which they tried out their new pedagogical ideas and then discussed their experiences on-line with their remote peers and with curriculum specialists and other "special guests" who joined the "virtual community".

The LabNet example is a type of networking which brings together a group of persons involved in the same educational task but who happen to be in different places. The PIT Schools Project in The Netherlands is another such example [15]. A different type of networking is "community networking", which often involve a wide ad hoc alliance of different groups from a single physical community--educators, community activists, local businessmen, parents, social service providers, government agencies, computer professionals, youth--interacting as peers via the framework of a "community network service" [16]. This model of a citizen's-based, geographically delimited community communication and information system has taken hold in hundreds of locations in the US and in many locations in Europe, New Zealand, Australia, Japan, Canada, and many other countries [13, 16, 17]. The technical device of a local BBS, a type of communication configuration, supports many of these, although others have much more complex technological aspects and more and more offer internetworking, most often via a gateway to the Internet system. The participatory-medium aspect of these networks, and the natural involvement of teacher and students within them, are bringing new "social and political architectures" [16], new partnerships, new interaction patterns, new access to education and training opportunities, and a new sense of community meeting place. They may also be stimulating a new type of locus of information and community centrality which used to be a role the school played in the community, but more and more has already been lost to shopping centres and other centres of gravity.

Thus there is some likelihood to believe that the impulse of networking, human networking amplified by technology, is an impulse that will take root in society and not be a passing cliche. What will be the result of networking when it confronts the organizational inertia of the school?

The secondary school and networking: constraints or integration?
The potential of technical networking to bring "virtual communities" into the school, parallel to the face-to-face communities available in the school, has been predicted to be a major development in education that is already well underway [18, 19]. Teachers and school administrators are discovering the stimulation and benefits of being able to interact with a self-chosen group of their peers, at a time and place convenient to them via asynchronous communication, and are taking increasing advantage of these possibilities [13,19]. The engagement of students in distributed communities is consistently felt to be an enriching and stimulating learning experience [19].

The networked computer laboratory, for example, is not an appropriate environment for networking activities. Will networking make a difference in the secondary school? Will it trigger a new deployment of communications and information technology in the school?

The start will probably be modest and tangential, with teachers and students doing their networking outside of the regular class and even outside of the school. (This is the case now with the growth of community networking and with the current patterns of network access shown by teachers and students. Honey and Henrmquez, for example, found that the most-frequent hours in which US teachers made use of the Internet was between 23:00 and 24:00 [19]). Occasional networking episodes in schools will for students will at first be of the nature of "electronic field trips" for students, and thus peripheral, relative to tests and grades and other socially expected measures of performance. The sense of concern about the costs involved for students to become part of on-line communities in a meaningful sense will deter organizational acceptance. Because of this we predict that teachers and administrators will be first candidates for meaningful networking [13].

Networking: integration, in time

It is my firm belief, however, that networking will eventually be integrated into the fabric of secondary-school education, in a way that many other applications of information technology have not. Why?

One major reason is that it taps a basic human need and pleasure - to talk, to make contact, to feel acknowledged by a group. We have not touched these needs at all with most of our deployments of information technology in secondary schools. This, plus the feeling of control that people have when they communicate on-line - they can say what they want, they use the computer as an extension of their ideas - are, I believe, contributing to the explosive growth of technical networking for communication and community building world wide.

Networking is also an application of technology that brings benefits into the school without needing to be adapted to the school. The outlay is relatively simple (compared, for example, to a multi-media laboratory) and the engagement of the participant is intense. For students, communication and communication-technology handling skills are practised in a functional setting. Teachers can find new and interesting resources and obtain new ideas from their contacts with their other communities, and they can increase their range of whom they can turn to with questions and problems. They can set the time and frequency of their networking, they can easily drop out of unproductive groups and try new ones. For once, the impulses of society may be in tune with what is also perceived to be useful and attractive to teachers [6, 13]. As has been said:

> ...there are two things that have to happen before an idea catches on. The idea should be good and it should fit with the temper of the age. (Jawaharla Nehru)

Community networking may be such an idea. It certainly fits with the temper of the age, particularly the impulses related to communication technology and the "superhighway" and the "global village" ideas of being flexible in place and time. It will probably not, in itself, revolutionalize the school. But it could be an application of technology in the secondary school that brings a new wave of interest, applications, and users. It may be a societal impulse which also becomes an educational impulse.

FINALLY: BETTER SOFTWARE

There is still a major need for good, easy-to-use, software to better support on-line communication. A user interface with at least some of the file-management functionalities of computer conferencing needs to be freely available, easy to install, and not require all participants to first be part of a complex and costly conferencing system [20]. But this is a topic for another reflection.

REFERENCES

1. Mecklenburger, J. (1994) *The New Generation of American Schools.* Inventing Tomorrow's Schools, **4** (2), 18-20.

2. Collis, B., & Anderson, R. (1993) *International assessment of functional computer literacy.* Studies in Educational Evaluation, **19** (2), 213-232. (For a larger range of references to support the conclusions in this section as well as throughout the paper, consult the author. Space considerations only allow a small sample of references here).

3. Pelgrum, H., & Plomp, Tj. (1991) *The Use of Computers in Education Worldwide.* Oxford: Pergamon Press.

4. Muffoletto, R. (1994) *Technology and Restructuring: Constructing a Context.* Educational Technology, **34** (2), 24-28.

5. Cuban, L. (1986) *Teachers and Machines: The Classroom Uses of Technology Since 1920.* New York: Teachers College Press.

6. Collis, B., & de Vries, P. (1993) *The Emerging Trans-European Network for Education and Training: Guidelines for Decision Makers.* Report prepared under contract to the Task Force Human Resources, Education, Training and Youth, Commission of the European Community, Brussels. Report No, 92-001-NIT-109/NL (This report contains a summary of five years of Community initiatives in the area of applying telecommunications technologies to increase equity of educational opportunity and mobility to all Europeans).

7. Fitzsimmons, E. (1993) *Office of Science and Technology Policy.* Washington, DC: Executive Office of the President.

8. Kapor, M. (1991) *Civil liberties in Cyberspace.* Scientific American, **265** (3), 116-120.

9. Balle, F. (1990) *The Information Society, Schools, and the Media.* In M. Eraut (Ed.), Education and the Information Society: A Challenge for European Policy (pp. 79-93). London: Cassell Educational Press.

10. From *Investor's Business Daily*, 4/14/94, p. 4; a typical source of this sort of statement.

11. Romiszowski, A. (1990) *Shifting Paradigms in Education and Training: What is the Connection with Telecommunications?* Educational Technology and Training International, **27** (3), 233-237.

12. Donaldson, J. (1991) *Boundary articulation, domain determination, and organizational learning in distance education: Practical opportunities and research needs.* Distance Education Symposium: Selected Papers. Part 3. ACSDE Research Monograph. Number 9. University Park: PA: The American Center for the Study of Distance Education.

13. Collis, B., Veen, W. & de Vries, P. (1993) *Towards a Communication and Information System for Education in The Netherlands: The CISO Project.* The Hague: PTT Telecom.

14. Ruopp, R., Gal, S., Drayton, B., & Pfister, M. (1993) *LabNet: Toward a Community of Practice. Hillsdale, NJ:* Lawrence Erlbaum Associates.

15. Collis, B. (1994) *A Triple Innovation in The Netherlands: Supporting a New Curriculum with New Technologies through a New Kind of Strategy for Teacher Networking.* In press.

16. Schuler, D. (1994) *Community Networks: Building a New Participatory Medium.* Communications of the ACM, **37** (1), 39-51.

17. Communications Canada (1992) *The Information Society: New Media ... New Choices.* Ottawa: Minister of Supply and Services.

18. Dede, C. (1994) *Swimming in a Sea of Information: The Future of Education in an Information Age.* Presentation at the Global Village Schools National Conference, Atlanta.

19. Honey, M., & Henrmquez, M. (1993) *Telecommunications and K-12 Educators: Findings from a National Survey.* New York: Bank Street College of Education.

20. Collins, D., & Bostock, S. J. (1993) *Educational Effectiveness and the Computer Conferencing Interface.* Educational Training and Technology International, **30** (4), 334-342.

Betty Collis is a Canadian who since 1988 has been a member of the Faculty of Educational Science and Technology of the University of Twente in The Netherlands. Her PhD was in the area of measurement and evaluation of computer applications in education, and since the late 1970s she has been active in numerous international projects in this area. She is particularly interested in the many different and complex factors that influence the usability of a technology in education, ranging from those at the micro level, relating to the design of the instrumentation itself, to the macro level, at which social, cultural and political influences strongly shape the direction and nature of usage. Currently, she is focusing most directly on applications of and issues relating to communications technologies in education.

Short Papers

During the conference a number of short papers were presented, eight of which are included here. Each contributes further to the themes in the main papers.

Teachers roles are explored by Niess; teachers working in new interdisciplinary teams using IT found their pedagogical knowledge challenged, as they felt they had to re-learn their own subject and how to teach it in a new interdisciplinary way. Although challenged, this left them feeling uncomfortable and insecure. In contrast Bottino and Furinghetti consider the role of software tools to support existing teaching needs. They worked from the assumption that teachers will only exploit the potential of software tools if their use is embedded in existing pedagogic problems. Immediately, therefore, the issue of integration from within, or designing new integrated structures, returns.

Baron and Harrari studied how students perceived informatics in school. Students self-evaluation of their competencies with the computer reflected the frequency of their use in the classroom. Blaho and Kalas introduced an Animation Microworld into their classroom in an endeavour to stimulate cross-curricular activities. This introduced the pupils to software development techniques through the creation of animation and control of the objects they design themselves. Moller also wished to capitalise on pupils' exploration by designing software to enable students to make discoveries. The challenge has been to reduce or even eliminate the amount of prompts needed for them to chart their own discoveries through the software environment. Again, it appears that if the opportunities are created, learners themselves move with ease into an environment where they can explore and use IT for themselves.

Both the concerns of teachers and learners are considered by McDougall, who found that teacher support for the introduction of innovation, in this case laptops, was absolutely critical; after a few years most students appeared to take more responsibility for their work. Tella also focusses on the change process for both, where full communication technologies were introduced to school settings. There, teachers experienced an increased awareness of theoretical constructs and their corresponding teaching practice, while students crossed the barriers caused by conventional lesson structures. Finally, Aston suggests that only forceful government action, making compulsory the use of IT in the curriculum, can enable progress towards integration to be maintained.

1 The teacher's role in a new problem-centred, interdisciplinary approach

Margaret L. Niess

Oregon State University, USA

Keywords: problem solving, interdisciplinary, curriculum development, teaching methods, pedagogy

Introduction

Changes in the work place present educators with formidable challenges, including demands from business and industry that employees know how to use mathematics, science and technology collectively, as well as separately, to solve problems. Educators have been asked to both develop and adapt processes for teaching concepts in a problem-centred, interdisciplinary manner much as in the work place. Traditional curriculum is based on essential concepts of each discipline and those concepts determine the applications students experience. In a problem-centred, interdisciplinary curriculum, problems determine which concepts from each discipline are needed.

What about the teachers who are asked to change from a discipline-based to problem-centred, interdisciplinary approach involving science, mathematics and technology? This qualitative study examined teachers' views of their subject matter and teaching of that subject matter as well as the changes in the teacher's role. Twenty-four secondary teachers, formed into eight teams of three teachers (one mathematics, one science, and one technology from a single school) were selected. Each team experienced a four-week business internship, gaining work place knowledge and skills and investigating mathematics, science and technology skills, and the concepts and competencies to emphasize. They gathered information regarding applications to provide contexts for problems. Next, the teams spent four weeks developing a problem-centred, interdisciplinary curriculum. During the school year, they taught mathematics, science and technology in an interdisciplinary context.

Subject matter structures

Teachers' notions of science and mathematics was composed of two components: concepts and processes used to develop those concepts. Their views were cast in terms of traditional science and mathematics curricula and were influenced by how they learned and taught that subject.

However, the technology teachers had less well-established views of their discipline. Their views were cast within computer skills rather than general concepts and processes. They focussed on applications (word processing, spreadsheets) rather than general concepts about computers and computing. They focussed on problem solving as a process for producing application output rather than developing general concepts. This difference may be due to the infancy of the discipline (computer science) as opposed to the long-standing traditions of science and mathematics.

After teaching in a problem-centred, interdisciplinary manner, all of these teachers' conceptions of their subject matter remained, but integration was added: "I probably think more about integration of knowledge". As Hauslein and Good [1] maintained, these teachers were now concerned with interdisciplinary teaching and they viewed the subject matter in that context.

Teaching Structures

These teachers were experienced, expressing the teaching of their subject matter much as expert teachers do. They were concerned with teamwork, problem solving, and cooperative learning as essential elements in teaching. One teacher depicted the elements of teaching integrated science, mathematics and technology as if it was a scientific experiment. A flask is set on a tripod above a Bunsen burner. Mathematics, science and technology are the three legs of the stand on which the flask, representing the learner, rests. The teacher is the energy source (the flame) providing the motivation for learning. In the flask, liquid made from problem solving, co-operative learning, team work and brain-storming, bubbles. The resultant steam leaves the top of the flask - the learning. In this model, "the teacher directs the learning, applies motivation (the heat), and is a guide rather than a dictator."

The Teacher's role

All the teachers viewed their role in this approach as "a guide rather than a dictator" or "an architect helping the students to build the learning they need." Another indicated that "the students are doing a lot more of teaching each other ... they are teaching each other spreadsheets." Still another teacher indicated "This is a different style for students to learn. They need to learn they have responsibility. I need to learn to let them take responsibility."

This changed role presented difficulties for the teachers. In the traditional, discipline-based approach, teachers prepared lessons to teach pre-determined concepts. In this new approach, lesson preparation changed significantly.

> "We don't do a typical lesson plan. The students come to us and tell us what they need to know. In the beginning we prepared extensive plans in each of our subjects because we were certain the problem would require specific concepts. But, we found that the students wanted to go a different direction and we needed to prepare a completely different set of plans."

Challenges

The problem-centred, interdisciplinary approach presented new challenges for teachers. The interdisciplinary approach called for teams of teachers working together with students. The problem-centred nature required longer blocks of time for investigation. Some schools tried 90-minute class periods, others tried three-hour blocks, and some only allowed the three teachers to swap classes, rather than team teach.

The scheduling challenge was coupled with the challenge of finding time for preparing as a team. Teachers needed time to develop the curriculum, identify the problems and plan for connections; they indicated, "a full teaching schedule doesn't allow for curriculum development." Yet, "If the curriculum is not all planned and well integrated, it doesn't work very well."

Teacher knowledge, both content and pedagogical, posed another challenge. These teachers indicated they were uncomfortable dealing with ideas and concepts in other disciplines. They reported feeling uncomfortable in these teaching situations because, although they felt secure about their own subject matter, they were uncomfortable when stretching beyond the traditional limits. If teachers need to integrate science, mathematics and technology they must have knowledge not only of the interdependence but also of the breadth and depth of each field, both the concepts and the processes.

Finally, teacher's pedagogical content knowledge was challenged. They indicated they knew the "traditional ways of learning the subject matter," but they needed to learn their subject in an "interdisciplinary way." They needed to learn to teach in teams. They needed to learn new ways of working with students in this problem-centred, interdisciplinary approach. One teacher commented, "I need to learn when to let them get confused and when to help them." As Shulman & Colbert [2] suggested, teachers, without the essential content base, find it difficult to discuss content and focus students' thinking and have trouble providing appropriate feedback.

References

1. Hauslein, P. L. & Good, R. G. (1989) *Biology content cognitive structures of biology majors, biology teachers, and scientists.* Paper presented at the annual meeting of the National Association for Research in Science Teaching, San Francisco, CA.

2. Shulman, J., & Colbert, J. (1987) *The mentor teacher casebook,* ERIC Clearinghouse on Educational Management.

2   Teacher training, problems in mathematics teaching and the use of software tools

Rosa-Maria Bottino

Istituto per la Matematica Applicata di CNR
Genova, Italy

Fulvia Furinghetti

Dipartimento di Matematica, Università di Genova
Genova, Italy

Keywords: software, teacher education, teaching methods, classroom practice

Introduction
Our work is grounded on our experience with a group of secondary school mathematics teachers. We have worked with them for many years in research about the teaching and learning of maths and in projects concerned with educational innovations in classroom. We refer in particular to the secondary school level with students aged 14-18.

As evidence from our work with teachers and many studies have illustrated, the mathematics teaching at this school level presents some crucial problems which are related to both contents, for example the introduction to proof, and methodology, for example the role and weight to assign to algebraic manipulation.

The use of software tools has been often suggested, by researchers and developers as a possible solution to some of the problems encountered by the teachers. But even if these tools are available in school laboratories, they are not often used in classroom practice, or are used in ways which are not very significant in relation to teachers' needs. One of the crucial aspects of the process which leads to the effective integration of technology into the educational environment lies in the way in which teachers acquire knowledge about, say, a given software tool and of its possible use in their teaching.

In this paper we present the main features of a teacher training project, addressed to secondary school mathematics teachers, in which the problem of the use of software tools is considered. In the design of this training course we took into account some aspects which, on the basis of our experience [1] and of the work of other researchers in this field [4], we have singled out as important in order to actually affect classroom practice:

- the 'goodness' of a software tool is not to be judged on the basis of general parameters, but in relation to given didactic problems and teachers' needs;
- in the case of the choice of software tools teachers do not have a consolidated experience and are often dependent by external influences;
- teachers have to be involved actively not only in the phase of the experimentation in the classroom, but already in the phase of choosing the materials to be used and in planning the didactic itineraries;

Starting from these basic assumptions we are working on the problem of training in-service teachers by trying to establish a relation between the mathematics teaching problems and the possible 'solutions' offered by the use of didactic software. Our approach is to consider first the teachers' needs and then possible products. Usually software tools are presented pointing out their 'desirability' such as technical features, opportunities offered and user interfaces available, but the context in which they are to be used is often not considered.

Concern of the work
The target of the course is a group of 50 in-service secondary school mathematics teachers. The course consists of about 20 hours, distributed in 7 weekly lessons. Five teachers with a good experience in educational research participated also in the design of the course. These teachers have the role of co-ordinating the work and mediating the opinions of the researchers (us) and the teachers trained. During the preparatory meetings, we asked these teachers to consider the mathematics subjects they wish to take into consideration; they singled out algebra, analytical geometry, calculus, and euclidean geometry. Afterwards teachers were invited to

produce a schematic list of key-problems encountered in the teaching of the previous subjects that they would like to face; they singled out the following list:

- weight to give to literal computation and in general to routine activities;
- how to introduce and train students in proof;
- role of intuition in the approach to the problems;
- role to assign to problem solving;
- correct use of a formal language.

Afterwards teachers considered some software tools which were selected on the basis of their availability in the schools where they teach, their wide diffusion and their already existing use in school practice. The following software tools were in particular considered: MATHCAD, DERIVE, CABRI, MICROCALC, LAM. One of these, LAM, is software sold together with a maths textbook largely adopted in Italian secondary schools. We include in the list of software tools to be studied also a spreadsheet, LOTUS, which is often present in school laboratories. Then examples of possible uses of software tools in relation with the educational problems singled out have been selected.

We give an example of how we work. One of the topics which teachers have identify as more problematic in their teaching is the introduction of proof. As is well known, proof involves different aspects, such as intuition, the role of language, and the process of abstraction and generalization. Some software environments can be used to perform activities connected with the informal reasoning level which precedes that of formalization. MATHCAD, CABRI and MICROCALC can be used with this aim. Visualization is one of the features of these tools to be exploited: the work of training consists not only of illustrating the different possibilities of visualization but also of linking the discussion of these possibilities to recent findings of the research in maths education in this field (see, for example, [3]).

References

1. Bottino, R. M., Forcheri, P., Furinghetti, F. & Molfino, M. T., *Mathematics and computers in upper secondary school: an outlook on experiences and opportunities,* to appear in Graf, K.-D. (editors) Proceedings of ICME Working Group 17.

2. Bottino, M. R. & Furinghetti, F. (1994) *Teaching mathematics and using computers: links between teachers' beliefs in two different domains,* in: Ponte, J. P. & Matos, J. F. (eds.) Proceedings of PME XVIII (Lisbona), v.2, 112-119.

3. Eisenberg, T. & Dreyfus, T. (1989) *'Spatial visualization in the mathematics curriculum'*, Focus on learning problems in mathematics, **11** (1), 7-16.

4. Noss, R. & Hoyles, C. (1993) *'Bob - A suitable case for treatment?'*, Journal of curriculum Studies, v.25, n.3, 201-218.

3   School environments and students' opinions on informatics

Georges-Louis Baron
Michelle Harrari

Institut national de recherche pédagogique
Montrouge, France.

Keywords: social issues, learners, classroom practice, management

Introduction
As with many countries, France is being faced with problems of integrating information technology (IT) in secondary schools. There is now a consensus that it should be integrated in every subject. But, if some undeniable "success stories" can be told, current observations show that, except in technical and vocational subjects, and to a lesser extent in scientific subjects, the use of computers is still considered an innovation and, as such, is not fully integrated in every day classroom practice.

Some of the key problems affecting integration are linked to the speed of evolution of technology, to the marginal place devoted to it in the curricula (which, in France are defined at the national level) and to the issue of teacher training. Previous work [1] has shown that this last issue is particularly problematic. As in other countries [2] future French teachers have a high level of expectation regarding IT and low levels of initial competence, while teacher training institutions can't afford much support.

In the meantime, young students are very enthusiastic about popular applications like computer games and appear to show abilities that adults do not easily show. It was interesting to study the issue of young students' ideas about computers and informatics in relation to the school context.

Aims, scope and methodology of the study

An empirical study, supported by the French ministry of education (Direction des lycées et collèges), was led at the INRP in 1993 in order to get a snapshot of the opinions of students attending junior high schools.

In the first phase, semi-directive interviews with eighth and ninth graders obtained preliminary information. This was validated with a questionnaire submitted to 1855 students aged about 14, studying in 28 junior high schools. Our sample was divided into two groups. In the main one (n=1544, 24 schools), the schools were randomly drawn. In the second (n=311), four schools leading innovative work in the field of IT were chosen. In order to avoid bias, in every school, the questionnaire was submitted to every students of the studied grade. Follow up interviews were also organised with teachers and school principals.

The rates of return were high. However, the results given below are only indicators about a reality which we could not study in full detail because it is so fast moving.

Diffusion of computers

About half of the boys and one third of the girls, 40%, claimed to own a 'real' computer ('ordinateur' in French, as opposed to game-dedicated computerised devices named 'consoles'). A majority of these computers were rather obsolete or game oriented; PCs and Macintoshes representing only 20% of the computer owners. A clear correlation was found between the possession of such a computer and the social milieu, with upper middle class students having more computers, more up-to-date and powerful than with the lower classes.

Over three quarters of the students said they played computer games, the most quoted being action games. No significant correlation could be established between the intensity of game playing and school achievement.

The questionnaire asked which five words came to students' mind in response to 'computer'. The most popular theme was about the external parts of the computer-video display unit, keyboard, mouse ... Very few students quoted internal structural components such as processor, or memory. Games were also a popular topic, mainly quoted however by the computers' owners. The major difference between the two groups appeared connected to the 'software' theme, significantly more present in the experimental group (one third) than in the other (one fourth).

Computer use in school

Several series of questions focussed on how students perceived informatics in school, which uses they knew, and which preferences and competencies in information processing they had. On how students perceive computer uses in school, results confirm that 'technology', a compulsory subject for every children during the first four years of high school, whose syllabus explicitly encompasses information technology) is the only subject where computers may be considered as really integrated. Elsewhere, except in some classes of the principal group and in all the four experimental schools, rather scarce usage of computers was mentioned. To a large extent the students' general opinions about informatics appear similar to ideas now widely spread within society such as "the computer will be everywhere", "everybody will have to learn using them", "computers create unemployment".

Concerning the place of informatics in the classroom, significant differences were observed in students attitudes between classes using IT regularly and others. When regular practice can be organised, it appears that IT potential for learning is appreciated. Students manifest a better knowledge of software tools and more positive views about IT utility.

A group of questions asked students to rate their competencies regarding several information processing tasks including word processing, file handling and robot piloting. Students in the experimental group did not rate themselves significantly higher. But, contrary to the principal group where a correlation existed with the possession of a personal computer and hence with the social milieu, no such correlation could be established in the experimental group.

Word processing came first (three quarters of the students claimed to use it "well" or "very well". They were also a majority for "loading a file from a disk", but only one third for "copy files from a disk to another". In both groups, boys rated themselves higher than girls.

Discussion

If only a minority of students owned a computer at home, most of them appeared to have used one in school. A large majority deemed they could process texts, but had problems dealing with operations on immaterial objects, like when copying files. It might be that they have acquired practical abilities closely tied to the operation of specific software but have little knowledge of concepts. Students in the experimental schools were more conscious of the potentialities and limits of computers.

Moreover, the differences in self-evaluation depended much less from possession of a computer and hence from the social milieu. This shows that, when favourable school contexts exist, students do acquire some basic competencies in information processing. We could not identify which factors play a major part to create such contexts. However, we noticed, in follow-up interviews, the importance of school managers [3], who are key persons to facilitate the purchase of computers and to support teachers teams leading innovative pedagogical actions. Further research will certainly bring a wider perspective to this debate which has already begun [4].

References

1. Baron, G-L. and Bruillard, E. (1994) *Information technology, informatics and pre-service teacher training.* Journal of computer assisted learning **10**, 2-13.

2. Wild, P. and Hodgkinson, K. (1992) *IT capability in primary initial teacher training.* in: Journal of computer assisted learning, **8**, 78-89.

3. Baron, G-L. and Bruillard, E. (1994) *Towards the integration of information technology in compulsory education? Potentialities and Constraints.* Wright, J. and Benzie, D (eds) Proceedings of IFIP WG 3.5 Conference, Philadelphia.

4. Bright, G.W. and Waxman, H.C. (1993) *The future of research on technology and teacher education.* Waxman, H.C. and Bright, G.W. (eds.) Approaches to research on teacher education and technology, AACE, 115-120.

4   Animation microworld for children as developers

Andrej Blaho
Ivan Kalas

Department of Informatics Education
Comenius University, Slovak Republic

Keywords: curriculum development, creative, classroom, open learning, logo

Introduction

We are developing an experimental curriculum of informatics. It is a separate subject, having 2 hours per week for 12 years old students. Our conception an "informatics as a subject" approach; we would rather use information technologies as powerful tools for cross-curricular activities.

We have described the first year curriculum elsewhere [1]. Our main objective for the second year - inspired by [2] and by our visit to Hennigan School in Boston - is to let children become the developers of educational software. They are still only beginners in Logo - we have not yet spent much time on Logo programming. However, we believe that children will be much more motivated if they develop serious educational material, although the process of developing it is "the cream". For example, we want to equip them with the following mighty and complex software development techniques:

- how to create a screen animation of a certain object (a turtle, as far as we are in Logo environment), and how to develop its behaviour;
- how to control this object from the keyboard or by mouse;
- how to create several animated objects (turtles) which interact both with the environment (with the background) and with each other - for example, they change their behaviour whenever they bump into each other.

Animation microworld

Therefore we decided to start with an Animation Microworld (developed by us in Comenius Logo for Windows). Four goals are the main concern here:

- children should discover how important the techniques listed above are for any attractive modern educational software;

- their design abilities such as planning, defining the goal and means, creating an interface should develop considerably;
- they should experience a *collage* way of work. Animation Microworld is a kit of pre-defined bits of objects and pieces of behaviour. Students develop screen compositions, usually simple interactive multi-agent games, by combining these bits together. This is a promising approach taken even in serious university courses, see for example [3];
- they should learn several new features of Logo programming itself, which they will soon need in latter developments. The Animation Microworld is a Logo project which intentionally keeps the Logo way of work; the bits of the kit are Logo procedures and the result of the development is a set of Logo procedures.

When children start the Animation Microworld, they are presented with a developer's control panel, namely a column of standard Windows buttons with predefined texts and actions, see figure 1. They click on the New Turtle button and choose an object out of the library of all objects which can be extended by anyone - imagine them choosing a Cat. When they press the Start button on the panel, they are pleased to discover that the Cat walks on the screen. Each object in the library brings with it its initial behaviour; children soon find out what happens when they click on the cat - it turns left and keeps on walking.

Beside its initial behaviour, each object brings with it a set of possible moves and events which it recognizes. When children press the Develop button, a table describing the Cat's current behaviour is displayed, see figure 2. Here

Figure 1

the object's starting position, heading, the way how it moves and the list of actual events, can be observed and modified.

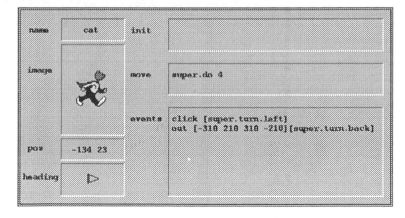

Figure 2

Let us click the Events sub-window. From the list of Available Events templates we can, for example, compose a new event, see figure 3:

key "S [super.will.footprint]

which means that whenever we press S, the Cat will start to leave footprints. In the similar way we can add some other reactions to other keys: the Cat can fly, jump, stop, fall, or change its behaviour if it steps on a certain colour. The Cat's animation is performed by a sequence of frames.

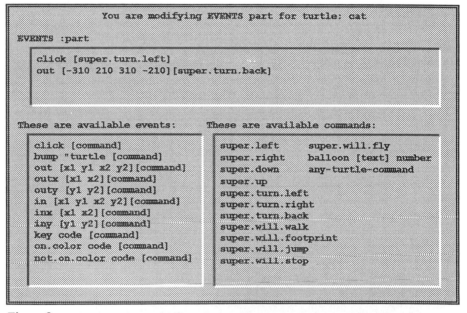

Figure 3

More complex behaviour can be designed if we create more objects, turtles on the screen. Our students explore this by creating a train, an engine with one or more cars. The problem occurs, however, if we click on the engine - it turns slightly right but the cars don't. To improve this, the way how the first car moves, for example, must be set to:

follow "engine 5

where 5 is the speed of the engine. If we change this number, the car can fall behind or get ahead of the engine.

After getting familiar with the Animation Microworld children start to develop more complex compositions. Figure 4 illustrates a game in which there is a Cat kept in the cage. It angrily walks all around and tries to reach the key by a ball. When the ball hits the key, the cage opens and the Cat walks out of it. The realization is as follows: whenever the Cat steps

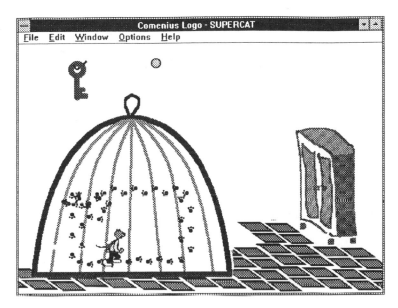

Figure 4

on the dark blue colour (the outline of the cage), it turns back - therefore the Cat cannot run out of it. The ball is moving on the screen, its direction can be slightly changed by clicking a mouse or pressing a key (depends on the decision of the designer). If it bumps into the key, another hidden turtle opens the cage, so the Cat can now walk out of it.

The Logo nature of the Animation Microworld is quite important for us; the complete development work is constantly and automatically being transformed into the four Logo procedures Screen, Turtles, Moves and Events - they correspond with four bottom most buttons of the control panel and can be directly edited by clicking these buttons. Our students have quite soon started to use this direct option for some kinds of modifications.

Beside accomplishing the four goals listed above, the Animation Microworld - being realized completely in Logo - confirms our metaphor of Comenius Logo for Windows as an Environment for Environments [4].

References

1. Blaho, A., Kalas, I., Matusova, M.: Experimental Curriculum of Informatics for Children Aged 11, submitted to IFIP WCCE'95, Birmingham.

2. Harel, I., Papert, S. (1990) *Software Design as a Learning Environment,* in: Harel, I. (ed.) Constructionist Learning, MIT Media Lab 19 - 50

3. Brown, C., Fell, H. J., Proulx, V. K., Rasala, R. (1992) *Using Visual Feedback and Model Programs in Introductory Computer Science.* Journal of Computing in Higher Education **4** (1), 3 - 26

4.   Blaho, A., Kalas, I., and Matusova, M. (1994) *Environment for Environments: New Metaphore for Logo.* Proceedings of IFIP Working Conference Exploring a New Partnership: Children, Teachers and Technology, Philadelphia.

5   How to develop "discoveries"

Herbert Möller

University of Münster, Germany

Keywords: learning models, creativity, future developments

Introduction
Discoveries represent the highest level of human intellectual work. Therefore it is important to find out how the ability to discover can be taught, learned and trained.   Up to now the situation at school is not satisfactory.   Although the method of learning by discovery is considered valuable, teachers rarely use it in classes because they have difficulty leading the pupils by questions, hints and a general stimulation of interest. Further obstacles are the science-oriented curricula, lack of time and disciplinary problems.

At the present time there are reasons to discuss once again the pedagogical advantages of discoveries:

- economic competition requires more creativity;
- people have to be prepared for higher intellectual tasks which cannot be done by computers;
- information technology offers new possibilities such as simulation, hypertext, geometric tools and computer algebra systems which may be used individually and interactively.

Since every scientific result has been discovered once, it seems possible and desirable to develop ways which enable interested people to rediscover selected results. This paper gives a very short report of several projects concerning this objective, based in the Heinrich-BehnkeSeminar für Didaktik der Mathematik at the University of Münster.

Method and criterion

Since most people need individual help whenever they try to discover, we have created suitable programs to optimize the time and the possible success. The software development proceeds in three phases, namely, planning the steps of the discovery, organizing the screenplay and writing the program. The most difficult task is the path-finding, that is devising the routes the user will take with the minimum of prompts. The aim is to reduce or even eliminate prompts for the user, rather in the way an adventure game interface provides only minimum guidance. We have called this a question of "suggestivity". In many cases the solution can only be found by research which includes new derivations and proofs. Sometimes even new scientific results are needed as in the third of the examples given below. We call this method "recycling of discoveries".

The computer programs offer the working environment and the information chain gives impulses by questions, hints, invitations and other tips. It is left to the user to decide at any time how suggestive the information should' be. Most programs are only planned, but the theoretical background is available for any of the following examples.

Examples

i) *The strategy of the game of "nim"*

Two players put an arbitrary number of small things into heaps and then alternately take off pieces from one heap at a time. The one who takes the last piece wins. The strategy of this old game can be discovered by playing against the computer. The development consists of five steps. In order to receive an even low level of "suggestivity" we had to solve two problems, namely, the introduction of the arithmetic XOR-composition by playing with three heaps and the discovery of a "minimum rule" which uncouples the further steps from playing.

ii) *The Fermat-Torricelli point*

Numerous geometric results can be rediscovered by experimenting with the aid of powerful geometry programs (for example, CABRI). But critical situations may also occur. In the case of the Fermat-Torricelli point, which has a minimal sum of distances to the vertices of a triangle, it is easy to conjecture the result. But the only known elementary proof is based on the fact that the shortest polygon between two fixed points is the straight line. Since this idea is not obvious and cannot be suggested, the computer program offers at first two problems which introduce this issue and which give the experience of solving the more complicated task.

iii) *Planetary motion*

Information technology enables the user to select all needed previous knowledge. The main method is simulation of motions that have become discrete in central force fields and in gravitational fields on screen. A long standing missing link in the developmental work of finding the lowest level of "suggestivity" or prompts, has been filled in this case by a new result on conic sections which relates the tangent lines with a directrix.

iv) *Geometric tools in elementary analysis*

The conception called "elementary analysis", which is based on simplified concepts of limits, differentiation and integration, facilitates the universal use of computer teaching analysis. The geometric strand with about thirty "associative sequences of figures" contains all important results discovered with the aid of only five "geometric tools". Since the development of discoveries in analysis greatly depends on the underlying conception, it is important that the transition to classical analysis can be achieved by discovering the general notions of convergence, continuity and differentiability as examples. The simplified concepts help to save a lot of time. Therefore the method of learning by discoveries may also be used in classes. In this case teachers may prepare their lessons with the aid of the programs on the low "suggestivity" level.

v) *Algorithmic linear algebra*

Since we based linear algebra on a sequence of about thirty algorithms, it has become the most exciting field of discoveries for students. The corresponding programs not only give more insight by examples but support conjecturing better than existing computer algebra systems. The further developmental work will make use of a forthcoming book entitled "Algorithmic Linear Algebra" which contains many exercises marked "treasure trove" for encouraging far-reaching discovery.

vi) *A number-theoretical language*

Number theory still offers a rewarding experience in discovering. Therefore we have designed the high level language NTL1 (Number Theoretical Language 1) to handle high precision arithmetics. The user will come to appreciate the possibilities of establishing problems in a programming environment which leans on the necessities of mathematics and a rich collection of already implemented algorithms. Selected discoveries will be developed in modules constituting easily accessible libraries.

Conclusion

The ability to discover is considered valuable in learning and in practice. Information technology offers a lot of new possibilities to support discoveries. Learners may be guided by computer programs which provide a working environment and a mentor at the same time. In many cases it is a challenging task to develop sequences of proposals with a low level of "suggestivity".

6 Integration of portable computers into Australian classrooms: issues and outcomes

Anne McDougall

Monash University
Victoria, Australia

Keywords: classroom practice, curriculum policies, integration, learner centred learning, logo

Introduction

The first major initiatives in the use of portable "laptop" computers in secondary schools in Australia began in 1990, with projects in some private girls' schools in Victoria [1,2] and a primary-secondary school pair in Queensland [3]. Now there are many such initiatives throughout the country (see for example [4, 5]), and a number of these projects have been running for sufficient time to enable some useful evaluative comments to be made. This short paper outlines the issues that this experience has shown to be important, and summarises outcomes reported from these initiatives.

Funding

In most cases the school has provided the portable computers for students to use, often in class sets for borrowing for limited times, and sometimes for longer term individual use. In some of the more elite private schools parents are required or recommended to purchase or lease machines of particular types for their children to use at school and at home.

As well as purchase of the actual machines, funding is also needed for technical support, facilities for charging batteries and for whole-class display, and for secure storage, repairs and possible upgrading of the machines. The costs of these initiatives, and the fact that they are being taken predominantly in the private school sector in Australia, raise some serious questions of equity.

Practical Problems

Problems that have been reported include difficulties obtaining printouts, batteries running low in class, screen breakages during transport, students forgetting to save files, students not having computers because they have forgotten them or they are being repaired, and student complaints about the weight of the portable machines. Most of these seem to have been overcome in the first semester of the projects.

Teacher Support

All the projects have been associated with at least some levels of staff anxiety, predominantly due to inadequate articulation of purposes or to technical problems, or of stress due to the increased workload and amount of change that seems inevitably to accompany this type of innovation.

Teacher support for these innovations is absolutely critical. This can include subsidising teacher purchase of their own portable computers, clear articulation of the purpose and philosophy of the initiative, provision of time to work with the new equipment and software before using it in classrooms, provision of classroom support for technical problems, attendance at workshops, courses and conferences, and facilitation of development of networks of people with expertise available to help.

Parents

Significant initial parent opposition to use of portable computers was reported in one case [1], where the parents were required to provide the computers for their children. Other reported sources of parental concern are the possibility of deterioration in handwriting, that time previously spent on other subjects will be sacrificed to make way for computer activities, potential effects on students' vision, that the computers will be dropped, damaged or left on trams, trains or buses, that carrying the weight of the machines in addition to other school materials puts undue stress on young bodies, security of the machines at home, ergonomic considerations, that computers might induce epileptic fits, and that parents will no longer be able to assist students with school work as the parents do not know enough about the computers [5].

Generally however parents have been most supportive of these initiatives, as they expect a relationship between increased school computing and workplace success.

Curriculum

Several of the portable computer initiatives in this country have deliberately involved change in the existing curriculum and in teaching practices; some, see for example [1, 3, 5], have made extensive use of LogoWriter programming to this end. With LogoWriter, computers have been used as intellectual tools, suggesting implications for curriculum design, redefinition of curriculum goals, and exploration of non-disciplinary approaches to secondary school curriculum, with commensurate changes in assessment and reporting procedures [3]. This is a particularly interesting development in the light of the more vocational approach implied by parental, and in some cases governmental, expectations.

Outcomes

The Queensland Education Department project has been thoroughly evaluated as it proceeded, with separate reports being published at the end of its first, second and third years; these are described elsewhere [3]. After the first year there was substantial evidence to suggest that students had become more willing risk takers, demonstrated more flexible approaches to solving classroom related problems, and established a network of cooperative practices. Increased cooperative learning and risk taking have been observed in other projects as well [1]. However, despite intensive and committed efforts by the teachers, the tasks of constructing and supporting a learning culture, attaining technical knowledge and implications of its use, and choosing models, metaphors and practices as starting points for students remained elusive [3].

After the second year students in the Queensland project were reported to have improved in knowledge and skills related to computing. However some students, a minority, were unhappy about the prospect of a third year in the project, feeling that they were missing out on some learning and social activities. Although some students had become excellent programmers, many were not using the higher level thinking skills that Logo can sustain; this was attributed to the peer teaching approach used and the lack of modelling of advanced programming skills. This result might be found more widely, as other projects also report greater use of peer learning [1]. Some gender differences became evident in attitudes, motivation and computer achievement [3].

After the third year substantial changes in classroom organisation and management, teaching approaches and relationships with students were reported by teachers, and collaborative work had increased further. Teachers believed that most students had taken more responsibility for their learning, although not all students had achieved improved outcomes. Students were more confident with risks and challenges, and had improved research skills. Parents cited students enthusiastically doing homework, projects, assignments and research. This enthusiasm and sustained attention to tasks has also been reported by other projects [1]. Marked gender differences in attitude appeared in the Queensland project classrooms, with a substantial number of Year 8 girls declaring a strong dislike of programming.

Conclusion

Many of the portable computer projects have not had extensive formal or public evaluation. However a reading of such material that is available, and discussions with teachers and others involved in many of the projects indicate that some exciting student outcomes are occurring. They indicate that as experience is gained and shared, many of the difficulties encountered by the pioneers can be overcome, that some interesting developments in curriculum might be sponsored by these changes, and that a cautious watch should be kept on issues of gender and equity.

References

1. Grasso, I. and Fallshaw, M. (1993) *Reflections of a Learning Community: Views on the Introduction of Laptops at MLC.* Methodist Ladies' College, Kew, Victoria, Australia.

2. de Figueiredo, J. (1991) *Selection, Installation and Use of Laptop Computers.* In McDougall, A. (ed.) Proceedings of the Computers in Education Group of Victoria Conference. C.E.G.V., 86-94.

3. Finger, G. (1992) *Integrating Learning Technology in Queensland State Schools.* Department of Education, Queensland, Australia.

4. Newhouse, P. (1994) *Opening the Floodgates to Computer Saturation.* In Ryan, M. (ed.) Proceedings of the Asia Pacific Information Technology in Training and Education Conference. Australian Council for Computers in Education, 329-35.

5. Tritton, L. and Dandie, B. (1994) Parents' Concerns About the Introduction of Notebook Computers into the School Setting: A Survey of Eight Private Schools. Unpublished paper, Monash University Faculty of Education.

7 Developing an open learning environment with communication technologies

Seppo Tella

University of Helsinki, Finland

Keywords: networks, electronic mail, telecommunication, classroom practice, learner centred learning

Introduction
Information technology was introduced into Finnish schools in the 1980s mainly through CAI/CAL packages and by using the computer as a tool. It was soon noticed, however, that the revolutionary influence of the computers across the curriculum was downgraded as schools turned IT to an autonomous school subject. A shift in emphasis took place in the early 1990s when careful attention was paid to computer-mediated communication, electronic mail, computer conferences and other telematic tools, which could help create an open, multimedia-focused and globally networked learning environment based on a constructivist conception of learning [1, 2].

The Utopia project
As a logical step onwards, a 2½-year long research project was launched in 1992 by the Finnish IT Centre for Schools, part of the Helsinki University Continuing Education Centre, in cooperation with the Helsinki University Department of Teacher Education. The project was mainly funded by the Finnish National Board of Education but also indirectly sponsored by the Finnish Academy. The project was called Utopia, an acronym for the Finnish words 'New Technology in Education-Pedagogically Innovative Thinking', comprising ten schools, comprehensive and senior secondary, in the Helsinki Metropolitan Area and involving about 120 teachers and headteachers with their pupils and students.

The project, in line with the qualitative research tradition, was designed as a developmental action research relying on collaboration between teachers, and aiming at developing an inter- and intra-school teaching and learning communications network. The methodological approach was focussed on evaluative, pragmatic and collaborative action research with ethnographic data gathering techniques. The formal part of the project was completed in mid-September 1994, though the social and telematic network formed between schools, teachers, the researcher, and a number of IT experts still continues at a very active level [3, 4].

Theoretical background

The topicality of the project was motivated through a number of changes - administrative and educational - in Finnish schools. Global networking was understood also to enhance human communication and gradually make up a new kind of telematic infrastructure.

Instead of merely transmitting information, teachers are generally argued to become many-sided consultants and facilitators, "guides on the side" instead of what they used to be, "sages on the stage". Teachers' visions, beliefs, and conceptions are also underscored when developing the schools and teachers' own work. Active learning and flexible uses of information sources are underlined in a technology-rich learning environment.

The conception of learning and knowledge is focussed from behaviorism on contextualism and constructivism [5]. International communications networks and various tools of modern technology contribute to creating a constructivist learning environment.

An open multimedia-based learning environment can be depicted as a symbiosis of a virtual, that is, telematic school and a physical school, where a number of levels of computer proficiency are possible. Telelogic communication has gained considerable ground at the expense of traditional monologic communication, leading to a number of changes in learning paradigms, learning and knowledge conceptions, IT applications, communication as well as in the status of teachers, students, and schools [3].

The research task and objectives

The research task of the Utopia project was to create an open, multimedia-based, network-focussed learning environment, whose development and study was one of the main problem areas. Research was also focussed on the change processes of the school itself, teaching practices as well as teachers' beliefs, preferences and attitudes. The participating schools were chosen through pragmatically informed and deliberate selection.

The background philosophy was based on the principle of ethogeny of the anthropomorphic model, and on symbolic interactionism. Ethnographic data gathering techniques included analysis of the communication transmitted via the common computer conferences, personal e-mailing, school consultations, participant observation and teachers' questionnaires.

Levels of expertise towards telematic proficiency

Enhancing teachers' and students' expertise in computer-mediated communication and telematics takes time and calls for a conscious effort. A good command of "telematic proficiency" can be divided into various levels of expertise. For instance, a fair command of e-mail and telecommunications software is based on a sufficient control of the man/machine interface of the computer but equally on the person's skill to use a word processor. We argue that beside e-mailing modern teachers of any subject need to master various telematic tools, such as Gophers, and, more recently, World Wide Web (WWW), Mosaic, etc. The levels of expertise towards telematic proficiency are depicted in Figure 1.

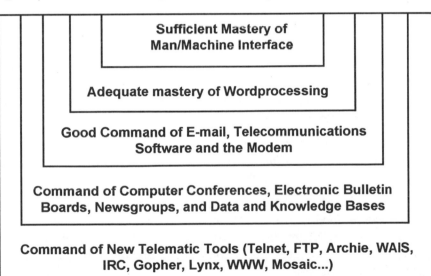

Figure 1: Levels of Expertise Towards Telematic Proficiency [3]

Results

One of the main results was a successful establishment of a constructivism-based, network-focussed learning environment, in which both teachers and students could master and control the structure of their own learning milieu and their own learning and studying processes to a certain extent, with the tools of modern communication and information technologies (CIT). The teachers experienced an increased awareness of the relation between certain theoretical constructs and the corresponding teaching practices. They also grew more professional users of CIT.

Telematic proficiency, the Utopia teachers advanced far beyond an "average" teacher. The students, when simulating a virtual school, easily crossed the barriers caused by conventional lesson structures and classrooms. A shared computer conference proved a fruitful concretion of virtual school concept, combining the participating schools and providing them with an electronic resource bank and a repository of ideas. The appreciation of the schools in their own municipalities increased thanks to the project, which also helped them have a modern profile among the rest of the schools in their municipality. Implementing CIT turned out feasible in teaching different school subjects, while the teaching-learning process could equally combine various teaching forms, such as collaborative learning, self-regulated learning, autonomous studying, and problem-based learning.

References

1. Tella, S. (1991) *Introducing International Communications, Networks and Electronic Mail into Foreign Language Classrooms: A Case Study in Finnish Senior Secondary Schools.* Department of Teacher Education. University of Helsinki. Research Report 95.

2. Tella, S. (1992) *Talking Shop via E-mail: A Thematic and Linguistic Analysis of Electronic Mail Communication.* Department of Teacher Education. University of Helsinki. Research Report 99.

3. Tella, S. (1994a) *Uusi tieto- ja viestintätekniikka avoimen oppimisympäristön kehittäjänä. Osa 1. (New Information and Communication Technology as a Change Agent of an Open Learning Environment. Part 1.)* Department of Teacher Education. University of Helsinki. Research Report 124. (In Finnish)

(4) Tella, S. (1994b) *Uusi tieto- ja viestintätekniikka avoimen oppimisympäristön kehittäjänä. Osa 2. (New Information and Communication Technology as a Change Agent of an Open Learning Environment. Part 2.)* Department of Teacher Education. University of Helsinki. Research Report 133. (In Finnish)

5. Duffy, T. M. & Jonassen, D. H. (eds.) (1992) *Constructivism and the Technology of Instruction.* Hillsdale, NJ: Lawrence Erlbaum.

8 The British approach to the integration of IT into the curriculum

Mike Aston

The Advisory Unit: Computers in Education
Hatfield, United Kingdom

Keywords: case studies, curriculum policies, integration, national policies, quality

Recent history

The UK's efforts in stimulating the use of IT in the curriculum have been well chronicled at international conferences and in academic journals since 1970 and at the 1st World Conference on Computers in Education, held in Amsterdam. The pace of change has been so great during the last few years [1], that little time has been devoted to reporting on the policies and strategies causing these changes. The U.K. Government's Education Reform Act of 1988 has prompted unprecedented and fundamental changes in the way in which education is delivered and assessed [2]. Information technology is firmly embodied in both the policy and the strategies that have ensued. The key features of the Act are:

- The legal status of a well-defined National Curriculum;
- National testing against specific criteria;
- Significant increases in autonomy of schools;
- Regular inspection of schools within a published national framework.

In essence, IT has a curriculum of its own (from age 5 to 16) with level descriptions and programmes of study [3]. IT is also expected to be used in both the teaching and learning of all subjects in the curriculum. National tests are intended to be carried out at certain ages. Schools may expand their technology resources without recourse to local or national government controls. The IT provision, along with all other aspects of the school is being evaluated every four years by independent inspection teams, who will report to the Office for Standards in Education, the school management and governors and the parents of children in the school. Teachers are only just beginning to come to terms with this new regime.

Convergence
The progress of the integration of IT into the curriculum can be measured by looking at the convergence and cohesion of four key elements:

- the curriculum
- the software
- the hardware

and most importantly,

- teacher education and professional development.

If any of these elements become out of phase with the other three, or worse still, two elements become detached from the time cycle of the other two, then nothing changes in the classroom or even regression may occur.

Different agencies are, of course, responsible for the progress of each of the four elements. In the UK, the curriculum is determined by the School Curriculum and Assessment Authority, influenced by a wide range of professional bodies. Software, apart from a relatively small number of government sponsored projects, emanates from private educational software houses and the commercial giants such as Microsoft. Hardware developments in schools are largely driven by market forces and the commercial interest in staying ahead of the field. Teacher education, in many ways, is the least well-provisioned element in the progress stakes.

The curriculum
Teaching and learning in schools clearly influence the development and subsequent marketing of software. The UK curriculum for Information Technology in England and Wales is delivered in two distinct ways. These two ways can be summarised as

- The nature of IT and the skills associated with its application;
- Using IT in the teaching and learning of the whole curriculum.

The first is well defined by programmes of study and level descriptions describing the types and range of performance which students working at a particular level should characteristically demonstrate. The second way is not so clearly defined but may be encapsulated in such statements as "where appropriate, pupils should be given opportunities to use IT to support and enhance their learning of mathematics". The words "where appropriate" are of little help to the teacher in the classroom. Guidance will need to be given, but it is highly likely to be non-statutory i.e. strictly for guidance only.

Evaluation

For the first time in the history of the English and Welsh Education System, a formal and standardised system for inspecting schools has been put in place under the 1988 Act. Teams of trained inspectors, contracted by the Office for Standards in Education, visit schools on a rotating basis every four years and report on all aspects of the activity in the school. The inspection is carried out according to an agreed Framework [4]. Schools, governors and parents all have access to the Framework and to the subsequent report. The inspectors focus on how the curriculum is delivered through the quality of teaching and learning in all areas. The teams are expected to have a specialist IT inspector and the various subject inspectors should have a supporting brief to look at the use of IT in their particular specialist field. All reports are summarised and stored in an electronic form for further research by such bodies as Her Majesty's Inspectorate.

Concerns

It would seem that all is in place for successful integration. The curriculum has been given a 5-year moratorium on change - this might be a problem in itself. Could it be too bland as a result? Software development is slowing down for a variety of reasons, not least among which are those concerning financial viability in a fragile market. The hardware manufacturers continue to surprise us with new platforms, interfaces, learning environments, often without any upwards compatibility. Teachers, with the best intentions, cannot find the time or resources to maintain their own confidence and competence with the technology.

In the UK currently, the use of knowledge bases on CD-ROM in primary schools, 100% access to portables by students, integrated learning systems and multimedia are being evaluated. Next year, there will be a different list, perhaps voice input, tapping into the global highway, personalised databases and so on ...

Conclusion

It could be argued that a good mix of "carrots and sticks" works reasonably well in Britain. Teachers like to have a good deal of autonomy but with some strong guidance and a framework within which ideas can be consolidated. We should be celebrating 25 years of remarkable progress.

References

1. Cabinet Office, Information Technology Advisory Panel (1986) *Learning to Live with IT.* HMSO, London.

2. HMI, Department of Education and Science (1989) *Information technology from 5 to 16.* HMSO, London.

3. School Curriculum and Assessment Authority (1994) *Information Technology in the National Curriculum.* Draft Proposals, HMSO, London.

4. Office of Her Majesty's Chief Inspector of Schools in England (1993) *Handbook for the Inspection of Schools.* HMSO, London.

Focus Group Reports

The programme of this Working Conference included focus group discussions of questions posed by the Programme Committee, with the aims of contributing to the debate during the conference and of preparing a position paper of conference views on key issues. Nine Focus Groups were organized, each considering one of four proposed themes, learning, curriculum practice, teaching and organisation; from these they issued recommendations built on participants' experiences and points of views.

The following reports are the outcome of these discussions which took place over four days of deliberations interleaved with other conference activities, including a Poster session presentation of preliminary ideas and a plenary Panel discussion of interim reports. It is hoped that these reports, while serving as a means of inspiration for the future work of IFIP WG 3.1, will contribute to the thinking and planning of policy makers at all levels, and help with the formulation of future strategy in support of integration of information technology into the educational world.

Focus Group: Integrating IT into learning

Participants
Walter Burke (GB), Bas Cartingy (NL), Val Clarke (AUS), Alison Griffith (USA), Ivan Kalas (SO), Claus Möbus (D), Misericordic Nomen (E), Anthony Parra (E), Hannah Perl (IL), Anne Rasse (F), Joy Teague (AUS)

Chair: David Johnson (GB)
Rapporteurs: Qi Chen (PRC), David Squires (GB)

Main issues
The group felt that the notion of *integration* needed to be clarified in this context. For example, it was noted that integration could be interpreted with respect to informatics education as well as informatics in school subjects. IT includes both concepts and tools which can be used to facilitate learning in the broadest sense, being dependent on the availability of resources and the manner in which the resources are employed. Within this context the main issues discussed were: perceptions of learning, the relationship to other media, motivation to use IT in learning, and learning away from the school.

Perceptions of learning

Learners will usually perceive the learning aims and context in their own terms, which may not be in those of the teacher or designer. For example, software is interpreted subjectively by the user, in this case, the learner. However, well defined software provides for multiple interpretations and multiple uses.

IT can be used to facilitate the transfer of control of the learning process from the teacher to the learner. Learners will learn when they choose the goals which are relevant to them.

A common feeling was that the use of software tools is very significant in an educational context. This significance is emphasised by the inclusion in some curricula of required project work which can be completed with the aid of IT tools. It was also noted that the completion of assignments with the aid of IT tools, such as DTP packages and graphics tools, also provides an impetus for students to use IT on a regular basis. The use of software tools allows an emphasis on skills, both at the level of knowledge acquisition and the development of meta-cognitive skills in an ever expanding spiral. This means that it is critical that recent developments in theories of learning should inform the design and use of educational software.

Relationship to other media

A group member commented "A school without books is only possible in a society without books. There are no signs that this kind of society is coming". It seems that in the foreseeable future the use of IT will need to be used in parallel with the use of more traditional media; a *multiple-media* approach as distinct from a multi-media approach in which different forms of media are interact through design as opposed to use.

Motivation to use IT in learning

The inclusion of computers in the assessment process provides a powerful motivation for learners to incorporate the use of IT in their work. The opportunity to use IT in the completion of project work provides a similar motivation, especially when learners discover that they can produce a more polished product and, in some instances, a saving of time and effort. Also, an IT based environment can be freer and more "fun". Students and teachers must be given the opportunity to recognise the benefits of IT.

Learning away from the school

Learning is not confined to the classroom or the conventional curriculum. IT has the potential for facilitating learning at home. One aspect of this is in distance learning of conventional curricula. A second area is where children choose to use IT to satisfy their personal needs. Two issues emerge:

- equity issues associated with home based learning;
- changes in perceptions of what constitutes appropriate learning.

If learning becomes predominantly or even partially home-based, what measures will need to be taken to ensure that all learners have the same access to hardware and software resources and local support? Is a new breed of peripatetic teachers envisaged, even if they only visit the home in an electronic form? Will the conventional perception of learning as something that has to be "worked at" be replaced by an "edutainment" perception? Will there be an even greater competition between the education system and the entertainment industry for the hearts and minds of children? In a home environment who will decide what constitutes appropriate programmes of study - learner, teacher, parent, ... What will be the role of telematics, for example, networking and electronic mail, in this context?

Focus Group: Curriculum practice 1

Participants
M. Angel De Miguel (E), Ichiro Kobayashi (J), Ard Hartsuijker (NL), Santiago Manrique Catalán (E), Herbert Moeller (D), Anton Reiter (A)

Chair: Peter Bollerslev (DK)
Rapporteurs: Yvonne Buettner-Ringier (CH), Zoraini Wati Abas (MAL)

Main issues
Many countries started IT education as a separate subject. Nowadays most countries integrate IT into the curriculum of other subjects. Some countries have a mixture of IT in education as a separate subject and integration into other subject areas.

IT integration varies between using computers as an educational tool and incorporating IT as new topics and new methods This differs from country to country and from subject to subject. The group discussed the successful integration of IT in Austria, Catalonia, Denmark and the Netherlands.

In order to integrate IT into the curriculum of school subjects, teachers and schools should have awareness, knowledge and a positive attitude towards IT and curriculum change. Teacher education should contribute in the realisation of the process of integration by:

- familiarising IT on the individual level and the school organisation level;
- activating the change in learning and teaching styles due to IT;
- influencing the revision of curriculum aiming at IT integration.

The focus group recommends that key decision makers and the general public be made aware of the following facilitating factors which will determine successful integration of IT into the curriculum:

- sufficient equipment;
- adequate teachers guidelines and teaching materials;
- suitable curriculum;
- qualified teachers and resource co-ordinators, consultants, technicians;
- resource centres;
- feedbacks from students, teachers, parents, schoolboard;
- incorporation into national syllabus and national examination program;
- authors integrating IT topics into their textbooks;
- international collaboration;
- vendor support.

The group is aware that, even within one country, there are great differences between planning and practice of the above mentioned facilitating factors. Constraints were also noticed such as financial limitations, overloaded curriculum and timetable, political interests and resistance to change.

In order to improve IT integration in many countries, the IFIP-UNESCO curriculum should be widely circulated to key decision makers and curriculum developers throughout the world.

Focus Group: Curriculum Practice 2

Participants
Doug Brown (GB), Ton Ellermeyer (NL), Marta Krausova (CZ), Walter Oberschelp (D), Jim Schultz (USA)

Chairman: Gail Marshall (USA)
Rapporteurs: Georges-Louis Baron (F), Barry Blakeley (GB)

Main issues
The group was conscious of interacting with a very difficult subject, as curriculum has manifold meanings for different people and in different countries. Several dimensions need to be considered such as the ideal, the published, the taught, the evaluated and the learned curriculum. In a sense we were exploring new territory, identifying milestones or possible significant features in a future curriculum map. It was like exploring new territory and we were struck by the knowledge that Barcelona was the home port for Columbus, who was the link between the old and the new worlds; and Catalunya was the country of Abraham Cresques who charted terrae incognitae in the fifteenth century.

Recommendations
We can present here only the bare bones of a rewarding and stimulating discussion over four days, giving the main recommendations and a few concrete references provided by the group members.

Assessment: When the use of IT is an essential part of students' application of their knowledge, IT must be present in both the formative and summative aspects of the assessment process.

The group noted that SAT examinations in the USA now carry a recommendation that students will be disadvantaged if they do not make use of IT, and that French technical/vocational subjects have integrated IT in assessment and also in teacher recruitment.

Balance and tension: The group maintained that balance and tension are unavoidable in curriculum development. Among dimensions that were considered were:

- on the one hand the speed of change of IT, and on the other the time needed for good curriculum development;
- on the one hand the changes created by IT and on the other the changes enabled/supported by IT. The group suggest that IFIP and WG3.1 should keep curriculum issues as a thread in all future conferences.

The use of Cabri-Geometre was sited as an example.

Diffusion: A key issue is to convince or persuade teachers of the demonstrable effects of technology use.   Hence the group identified the need for a system of dissemination of short synopses of programmes and models of integration, including failures as well as success stories.   Here we present the first ideas for structuring such synopses, which should encourage people to appreciate the most promising ideas for their particular situation.   In that respect, synopses should refer to fairly substantial works, involving at least one school and several teachers over a significant period of time.   The contents of synopses might include:

- full references, aims, contents/age/subject area, resources (including particular backup), size/duration, finance/resources (how and where obtained);
- what were the successes?
- what were the failures?
- what would you do differently?
- what resources were used successfully?

Focus Group: Teachers and Teaching

Participants
Giorgio Casadei (I), Steffen Friedrich (D), Kurt Kreith (USA), Colette Laborde (F), Manoli Pifarré Turmo (I), Andreas Schwill (D)

Chair: Paul Nicholson (AUS)
Rapporteur: Paul Zorn (USA)

Main issues
The general focus was how IT changes the processes of teaching and learning.   At first the group identified several very general roles that computers might play, including:

- amplification of thinking: providing such resources as "horsepower" and speed to the human mind;
- feedback, by monitoring processes and producing concrete outputs;
- helping students to acquire meta-cognitive skills, such as algorithmic thinking, information organization, knowledge transfer, process control;
- permitting multiple representations and manipulations of objects and ideas.

Key questions and some answers
What do teachers need to know about IT in order to gain benefits mentioned above? They should

- understand general ideas behind any form of programming, but not necessarily details of coding programs in particular languages;
- understand a variety of IT tools, including word processors, spreadsheets and hypertext, so that they and their students can choose intelligently, and also understand the need for new tools;
- participate in developing educationally appropriate software, by being able to evaluate IT tools both as software and as educational resources;
- know whether and how IT fundamentally changes their own subject areas, for example how CABRI changes ways of doing and thinking about geometry;
- understand the pedagogical and didactic processes that various IT tools aim to address;
- recognize with their students when computers produce nonsensical or useless answers - output isn't meaningful just because a computer produces it.

What can we learn from past experience? For example, in Germany, technology-based language laboratories failed. Among the lessons from this experience are that teachers should be deeply involved and appropriately prepared; also that tools should be used for specific intellectual purposes. From hand held calculators we learn that although such machines can relieve burdens of routine computation, teachers need to know what else to do and how to use the new freedom such tools offer.

Calculators can now store data, and may be used in French high school examinations; French students now use calculators to store information for use during examinations. This means that teachers must re-think what to examine, and students must organize carefully the information they decide to store. However, experience with earlier technologies is not fully relevant to IT, because computers are qualitatively different from earlier tools. The computer can completely rearrange the relationships among student, teacher, and subject.

What is the teacher's role in an integrated IT environment? One effect is that teachers need especially to know how to manage, structure, and manipulate information sources, for example through hypertext environments. Teachers change from dispensers of information to moderators, catalysts, organizers, and synthesizers. Another effect is that teachers will need at least as good a knowledge of their subject areas as before though perhaps somewhat different knowledge. In-service teachers may need special help in coping with new IT environments and new teacher roles, especially in a time of rapid change.

How can teacher expertise be "multiplied" efficiently? Sharing expertise effectively is difficult because teachers are used to working alone. Close collaboration must be built in to systems. Teachers from different disciplines might learn from each other by collaborating on a single general project, such as building a hypertext environment to describe many aspects of a town.

Focus Group: Managing integration 1

Participants
Magda Bruin (NL), Hugh Burkhardt (GB), Maria Cabezas (E), Klaus Graf (D), Paul Jansen (NL), Cassandra Murphy (USA), Nathalie Saulnier (F), Matti Sinko (SF)

Chair: Jim Ridgway (GB)
Rapporteurs: Pieter Hogenbirk (NL), Viera Proulx (USA)

Main issues
We address the issues of organization, management, and implementation strategies for integration of information technology into secondary education. We start by identifying the key arguments for using information technology in schools. These arguments can be used to support decision makers at all levels: politicians, educational administrators, higher education, including teacher training institutions, business partners and parents, as well as classroom teachers.

IT integration should concern ways to support teachers in their central role, not simply to take over part of it. IT integration is not an end in itself; it is essential to identify key curriculum roles for IT in different subjects, and to identify elements of the curriculum that will profit most from integration. All our actions should be driven by questions about the quality of the learning activities which will result.

IT has penetrated the world of work in many different ways and also entertainment. These changes represent an important source of ideas about how innovation in education might be managed, and these societal changes are, themselves, justification for educational change associated with IT, and justify getting resources for IT.

Principle level actors at all levels of the educational system need arguments they can use for integration of IT into schools. Some key points which can be used to convince politicians and administrators include:

- modern societies need to be increasingly 'smart' to survive economically;
- citizens who know nothing about technology are likely to become alienated in an increasingly technological society. This argument has important implications for access to IT by all groups, including those in economically deprived areas, and for females where activities may need to be designed to encourage them to use IT to the full;
- informed citizens need to understand the opportunities and threats which technology offers;
- the pressures for lifelong learning will require extensive use of IT;
- IFIP/UNESCO curriculum suggests we should do more;
- we need to prepare students for the workplace of the future;
- today, competency in using IT is as basic as reading;

To convince teachers to include IT in their teaching, several strategies can be employed:

- create a high quality IT based lessons that complement regular curriculum;
- include IT based activities into the existing 'high stakes' assessment, on which schools and teachers are judged, and provide curriculum materials needed to prepare for them;
- plan time and support for teachers' ongoing training and networking;
- create support centres with libraries of software and curriculum resources;
- aim to support weaker teachers.

Organisational factors
Politicians and administrators need to show their commitment to IT in policy statements that include specific targets to be attained over a period of time.

Co-ordination of IT activities at the school and district level should be done by teams that consist of:

- a senior teacher who can argue for resources with school management;
- an IT co-ordinator or teacher, knowledgeable in uses of IT in classrooms;
- a teacher who is an innovative teacher, but who may be IT naive.

Co-ordinators should have ways to assess success, and plans on how they will remediate failures which are certain to occur. Plans should be made for continued teacher development and networking that include initial training, mutual classroom visits, regular support group meetings, networking with peers and mentors, learning about new resources, and working on shared projects. Sufficient time should be allocated for these activities, possibly by reorganizing class organization or time structure.

Implementation level

Planning for IT integration should be done on a long term basis. Plan funding for several years ahead: for hardware, software, curriculum materials, technical support, and upgrades. Select curriculum activities that can be implemented with the projected funding and have teacher's support.

Look for alternative ways of using IT in the classroom: group rather than individual projects; one computer in classroom as a teacher assistant or demonstration tool; use of older equipment and software for lower level activities; use of computer as a tool on occasional basis.

Support teachers in developing their own IT competencies and professional growth through:

- home computer use via equipment loans or discounts on purchase;
- easy access to computers at school;
- providing vivid examples of uses of IT in classroom - including pedagogy, lessons, classroom visits, videos;
- involving faculty and graduate students at local universities as partners and mentors who suggest curricular activities, help with development of class materials and lessons, and who provide examples of applications of IT at the leading edge of their discipline, providing support and assistance with subject specific problems.

Focus Group: Managing integration 2

Participants
Lillian Cassel (USA), Jordi Castells Prims (E), Begoña Gros Salvat (E), John Mallatrat (GB), Mary Ann Robinson (USA), Josep Sales Rufi (E), John Scholtes (NL), Franz Stetter (D), Jacques Verwater (NL)

Chair: Margaret Niess (USA)
Rapporteurs: Mike Aston (GB), Ronald Ragsdale (CDN)

Main issues
Positive forces include skilled and enthusiastic teachers, successful models and dissemination, planning of innovation, demands of society and the creation and availability of software. Resistance emanates from reluctant teachers, short term views by governments, confusion between means and ends, curricula, schedules and examinations and the magnitude of the paradigm shift.

A model for change
The model proposed is learner centred and focusses on the school as the essential agent in the integration of technology in education. Although change needs to be effected within the school community, a recognition of strong links through subject disciplines with other local schools, and teachers, and even wider afield, needs to be amplified.

The school
The school requires a multi-faceted team led by teachers who are subject experts in the disciplines. They in turn are supported by the school principal, technical staff and curriculum assistants. Key features of the paradigm include the need to:

- plan new integration efforts based on the achievements and failures of successful models that are driving integration;
- utilize those skilled and enthusiastic teachers in the planning stages;
- provide reluctant teachers with opportunities to visit successful models, allowing time for reflection on their current pedagogical practices;
- have IT co-ordinators in the schools and provide them with good support and rewards;
- allocate finances in support of the sharing of ideas and experiences between teachers both within and outside the school and with other professionals within their disciplines;
- allow time for the organisational model to mature.

Outside school

Appropriate agencies need to:

- describe activities that IT makes possible and document the importance in education, identify benefits to the discipline that might come from these new activities, and support specialists who need to validate benefits and state clearly how they can be realised;
- create a climate that encourages teachers to innovate;
- support wide-spread dissemination of successful innovations;
- support development of model sites, managed by skilled and enthusiastic teachers, giving access to other teachers to observe and validate;
- establish computer network contact between teachers where appropriate;
- improve the quality of pre-service and in-service teacher development by integrating the use of technology into teachers' courses in their disciplines;
- make available a resource directory to include sites to visit and reports on experiences;
- establish education programmes specifically for co-ordinators, including development of group motivation skills;
- establish partnerships among various stakeholders for implementation, innovation, and dissemination.

Focus Group: Teachers as integrators

Participants

Miguel Angel Aguareles (E), Francesca Alloatti (I), Krystyna Dalek (PL), Brian Durell (CDN), Karoly Farkas (ROM), Helene Godinet (F), Joyce Currie Little (USA), Pere Marques Graells (E), Anne McDougall (AUS), David Miller (GB), Joan Berga Reixach (E), Takashi Sakamoto (J), Robert Smith (USA), Seppo Tella (SF)

Chair: Anna Kristjansdottir (IS)
Rapporteurs: Charles Duchateau (B), Lilliam Hurst (CH)

Main issues

Today's school, through its organisation and preoccupations, has built many barriers, resulting in an isolated school. This may be represented as a gap between society in one circle and school in another. IT is one component which falls into the gap between these two circles for the learner.

If school is to remain a preserved environment for learning, it also has to open new vistas and to integrate perspectives coming from life outside school. Integration is the effort to provide an intersection between the school and society. An integrator-teacher is a teacher seeking topics and means which will enable school-linked preoccupations to intersect with the outside world. This could be represented by a merging across the gap of the two circles, and where they intersect IT is a component of reconciliation.

IT is not a name, but a means enabling the reconciliation of schools and society as well as removing the barriers inside schools as well as between school and society. Efforts should be made to create an environment in which teacher attitudes and beliefs can change. The end result should be a skilled, enthusiastic, and self- and peer-supporting team of teachers.

Society
Schools and integrated teachers need three new "Rs" from society: Rewards, Resources and Recognition. Rewards do not only consist of financial ones but society should recognize their work and initiative by providing necessary resources enabling them to continue their work of integration.

School
The culture and structure of today's school render the job of integrator-teacher and also the integration of IT very difficult. From a short-term viewpoint, and because of school structure and organisational constraints, a way of using IT capabilities can consist of fitting into the present-day structure a new entity. This is no longer a computer science course, nor really any existing classical subject, but is performing learning activities because of and around IT. This new entity will be devoted to real problem-solving originating in the present day situation of the learners outside the school.

The education of teachers involved in these new activities must be carefully charted and teacher trainees must be provided with teaching scenarios and appropriate tools. So, this may not be an excuse for non-integration of IT in other subjects and practices, but an opportunity for the teacher trainee, and even for the unconvinced teachers, to become aware of what IT is and how it can be used.

Intersection between school and society

IT empowers average teachers to incorporate school activities into a virtual school concept which contains the symbiosis of a virtual school and the physical school, thus combining the virtual school with the real school.

Teacher's attitudes, beliefs and preference

Teachers have a central role in integrating IT into schools. They have to change their roles to become facilitators to help learners to use IT, which can be viewed as just another way to access knowledge. Thus, teachers' attitudes, beliefs and preferences have to be changed and adapted. We cannot fabricate this new teacher, only provide conditions for their emergence as well as support when they have emerged.

A good dissemination of IT is possible on the basis of one-to-one contacts and exchanges. The adoption of IT would also be made easier by integrating the need for using IT tools into existing curricula, and particularly by including IT features into assessment.

Focus Group: Teacher education 1

Participants

Norman Au (IIK), Norbert Breier (D), Terry Cannings (USA), Sou-yung Chiu (TW), Giovanna Gazzaniga (I), John Gardner (GB), Csilla Heves (H), Ferenc Iszaj (H), Eva Kluge (D), Margaret Nagy (H), Mihaly Nagy (H), John Olson (CDN)

Chair: Frank (Nick) Nicassio (USA)
Rapporteurs: Fulvia Furinghetti (I), Brent Robinson (GB)

Main issues

In order for information technologies to be integrated into practice, ultimately teachers cannot be obliged to change but must instead come to an understanding of, and willingness to, alter their current practice. It is important therefore that any programme of change, personal or mandated from outside, includes a teacher education programme, and that this programme begins from, and respects, teachers' existing beliefs, assumptions, expectations, attitudes and needs.

This applies to pre-service teachers as well as to serving teachers, though the nature of this mental set will vary between the two groups and of course, between individuals.

Proposals for technological education for teachers should therefore identify, acknowledge and build upon teachers' images of:

- the nature of their subject discipline;
- the teaching of that discipline;
- the technology.

Proposals for technological education for teachers must also relate to practical issues:

- pressure of time;
- the structure, administration and culture of the school;
- political compliance at a micro and macro level;
- expectations and needs generated by the wider society.

Often if we analyse the results of the implementation of externally motivated innovations, we cannot recognise the original intentions of the proposers. This happens because in the middle of this process there are teachers acting as filters between pedagogic proposals and classroom realities.

There is likely to be an inherent tension between:

- teacher educators' assumptions which are visions of what schools ought to be; and
- teachers' assumptions which are predicated upon a view of schools as they actually are or have been.

Teacher education should therefore include strategies for working with existing resources while fostering longer term vision.

It should also provoke the tensions which result from the interaction of the two and help teachers understand and cope with the nature of the change process, especially the move from a stable to an unstable state and the tensions and anxieties which may be felt.

Teachers need to be given concrete models of what can be achieved in terms of software applications, classroom activities and lesson planning. Teacher educators should:

- provide examples of effective classroom use of IT; and
- model the use in IT in their own teaching.

Teachers also need images of possibilities of what might be achieved. They need strategies for helping them to realise these possibilities.

As part of their technological education, teachers should be encouraged to form collaborative groups in order to:

- reflect upon and develop their beliefs, assumptions, expectations, attitudes and needs in a dialectical process;
- generate action;
- provide mutual support to members in their role as change agents.

Teacher education for change and innovation also requires the education of policy makers. The principles and guide-lines in this paper must also be applied to this other important group if effective technological innovation is to succeed in education.

Action priorities

IFIP should identify, through a review of the literature and the promotion of research, what changes might be expected in classroom practice and of these changes, which exemplars should be actively pursued.

IFIP should also take steps to identify the strategies by which these desirable changes can be effected by teachers.

Focus Group: Teacher education 2

Participants

Anton van Aert (NL), Joan Escué Olivé (E), Akira Horiuchi (J), Dirk Janssens (B), M. Dolores Santandreu (E), Sigrid Schubert (D), Shigera Tsuyuki (USA), Joerg Vogel (D), Rado Wechtersbach (SO)

Chair: Wim Veen (NL)
Rapporteurs: Erich Neuwirth (A), Anthony Ralston (USA)

Main issues

The proper and increasing use of IT in secondary schools (and at other educational levels) depends upon a symbiosis between the supply side:

- innovators
- national programs
- researchers
- software developers;

and the demand side:

- schools
- teachers
- students/parents

At present the demand side is rather weakly developed, as also is the national program aspect of the supply side. It is important that this weak development becomes stronger. This motivates the following general recommendations.

General Recommendation 1: In order to promote IT integration in education, appropriate steps need to be taken to stimulate demand for IT from teachers and schools and, as far as possible, also from parents and teachers.

This recommendation is echoed below in specific recommendations. On a somewhat less general level it is useful to delineate the various actors in the educational community:

- policy makers
- curriculum makers
- teacher educators
- schools/teachers
- students/parents.

Stronger links need to be established between various pairs of these actors. Of particular importance to this group is the link between teacher educators and schools/teachers. This leads to the second general recommendation.

General recommendation 2: Integrating IT into education requires close cooperation between teacher training institutions and the schools and teachers themselves.

Recommendations for plans for action
Teacher training institutions are playing a crucial role for any change in the educational system. So successful IT integration, in the long run, is simply impossible without a strategy for these institutions. In particular, the following factors have to be taken into consideration.

Partnership: Since IT integration is an ongoing process, not only teachers have to be trained, but feedback from schools to teacher training institutions can help to adapt the strategies of these institutions to the immediate needs of schools. Therefore partnerships between schools and teacher training institutions should be encouraged. Such partnerships should be fostered throughout the whole educational system.

Training the teacher trainers: Currently the IT awareness of teacher trainers seems to be the weakest spot with respect to IT in the educational system. Teacher trainers in general do not differ from school teachers as far as their beliefs and attitudes related to IT are concerned. Since with school teachers offering experience with computers as a personal tool enhancing productivity prove to be a main ingredient for successful IT integration in schools the same strategy should be applied for teacher trainers and teacher training institutions. Who should train those trainers? As in most teacher training institutions there are subject experts additionally having IT knowledge, peer learning seems to be the most viable strategy.

Train the student teachers: Student teachers should have experiences with IT during university studies in the following three different contexts:

- university departments should provide IT experience as part of the subject oriented studies;
- teacher educators should provide experiences with IT as a learning and teaching tool during the courses as well as during the internship;
- students should use IT as a personal productivity enhancement tool during their studies.

Train school teachers through partnerships: Ongoing in-service teacher training is a key issue for successful IT integration. As mentioned previously these training activities could be implemented within the framework of partnerships with teacher training institutions.

Involving teacher training institutions in policy activities: Teacher training institutions not only have the obvious task of training teachers, they also need to be involved in influencing curriculum changes software development creating supplementary educational material research.

Addresses of contributors and Programme Committee

ABAS, Zoraini Wati
University of Malaysia
59100 Kuala Lumpar
Malaysia
zwabas@zoraini.pc.my

ASTON, Mike
The Advisory Unit
Computers in Education
126 Great North Road
Hatfield AL9 5JZ
United Kingdom
mike@advunit.demon.co.uk

BARON, Georges-Louis
INRP
91 Rue G.Peri
92120 Montrouge
France
baron-gl@citi2.fr

CHEN, Qi
Department of Psychology
Beijing Normal University
Beijing
100875
PR China
Fax: 861 201 3929

CLARK, Valerie
Deakin University
School of Psychology
Geelong
Victoria 3217
Australia
vac@deakin.edu.au

COLLIS, Betty
PO Box 217
Univ of Twente
7500 AE Enschedde
The Netherlands
collis@edte.utwente.nl

CORNU, Bernard
IUFM Grenoble
30 Avenue Marcelin
Berthelot
38100 Grenoble
France
cornu@grenet.fr

DUCHATEAU, Charles
CEFIS
Facultes Universitaires N-D
de la Paix
Rue de Bruxelles, 61
8-5000 Namur
Belgium
cduchateau@cc.fundp.ac.be

FURINGHETTI, Fulvia
Dipartimento di matematica
Universita di Genova
Via L.B.Alberti 4
16132 Genova
Italy
furinghe@dima.unige.it

GRAF, Klaus-D
Takustr 9
Freie Universitat Berlin
14195 Berlin
Germany
graf@inf.fu-berlin.de

GRIFFITH, Alison
Department of Educational Leader-
ship, Counseling and Foundations
University of New Orleans
New Orleans, LA 70122
USA
aigel@jazz.ucc.uno.edu

HODGSON, Bernard
Departement de mathematique
et de statistique
Universit Laval
Quebec G1K 7P4
Canada
bhodgson@mat.ulaval.ca

HURST, Lilliam
CIP
PO Box 3144
1211 Geneva 3
Switzerland
hurst@cui.unige.ch

JOHNSON, David
CES, King's College London
Waterloo Road
London SE1 8TX
United Kingdom
UDUE040@uk.ac.kcl.cc.bay

KALAS, Ivan
Dept. of Informatics Education
Comenius University
842 15 Bratislava
Slovakia
kalas@fmph.uniba.sh

KOBAYASHI, Ichiro
7-5-10 Musashidai
Hidaka-shi
Saitama-ken
Tokyo 350-12
Japan
T2031.1362@compuserve.com

LABORDE, Colette et Jean-Marie
IMAG-LSD2
Universit Joseph Fourier
BP53X
38041 Grenoble
France
colctte.laborde@imag.fr
jean-marie.laborde@imag.fa

McDOUGALL, Anne
Faculty of Education
Monash University
Clayton
Victoria 3168
Australia
AnneMcD@
vaxc.cc.monash.edu.au

MOLLER, Herbert
Mathematics Institute
University of Munster
Einsteinstr 62
48149 Munster
Germany
mollerh@math.uni-muenster.de

MOREL, Raymond
CIP
PO Box 3144
1211 Geneva 3
Switzerland
morel@uni2a.unige.ch

MURPHY, Cassandra
Office of Research
Tulane University
New Orleans, LA 70118
USA
murphy@mailhost.tes.tulane.edu

NICASSIO, Frank
2001 SW Nye Avenue
Umatilla Education Service District
Pendleton, OR 97801
USA
nick-nicassio@umesd.k12.or.us

NICHOLSON, Paul
662 Blackburn Road
Deakin University
3168 Melbourne
Australia
pauln@deakin.edu.au

NIESS, Margaret
Science and Mathematics Education
Oregon State University
Corvallis, OR 97331
USA
niessm@ucs.orst.edu

OBERSCHELP, Walter
RWTH Aachen Math Institute
Ahornstr 55
Aachen 52074
Germany
angin@informatik.rwth-aachen.de

OLSON, John
Faculty of Education
Queens University
Kingston, Ontario K7L 3N6
Canada
olson@educ.queensu.ca

PASSEY, Don
Dept of Psychology
University of Lancaster
Lancaster LA1 4YF
United Kingdom
d.passey@centl.lancs.ac.uk

RIDGWAY, Jim
Dept of Psychology
University of Lancaster
Lancaster LA1 4YF
United Kingdom
j.ridgway@centl.lancs.ac.uk

RUIZ I TARRAGO, Ferran
Generalitat de Catalunya
Departament d'Ensenyament
Via Laietana 64, 1er
08003 Barcelona
Spain
ferran@pia.xtec.es

SAKAMOTO, Takashi
2-19-23 Komaba Meguro-ku
National Centre for University
Entrance Examinations
153 Tokyo
Japan
Fax: 81 3 5478 1290

SAN JOS, Carlos
NTIC, Ministerio de
Educacion y Ciencia
Torrelaguna 58
E 28027 Madrid
Spain

TAYLOR, Harriet
Dept of Computer
Louisiana State University
Baton Rouge, LA 70803
USA
taylor@bit.csc.isu.edu

TEAGUE, Joy
School of Computing and
Mathematics
Deakin University
Geelong, 3217
Australia
joy@deakin.edu.au

TELLA, Seppo
University of Helsinki
Hietalahdenk 16-18 D 101
00180 Helsinki
Finland
tella@cc.helsinki.fi

TINSLEY, David
Holly House
South Church Street
Bakewell,
Derbyshire DE45 1FD
United Kingdom

VAN WEERT, Tom
School of Informatics
University of Nijmegen
PO Box 9010
6500 GL Nijmegen
The Netherlands
school@cs.kun.nl

VEEN, Wim
IVLOS Institute of Education
Utrecht University
PO Box 80127
3508 TC Utrecht
The Netherlands
w.veen@ivlos.ruu.nl

WATSON, Deryn
CES, King's College London
Waterloo Road
London SE1 8TX
United Kingdom
deryn@bay.cc.kcl.ac.uk

Index of contributors

Keyword index